To Michelle
With love,
Christmas '80 —
Aunt Bonnie

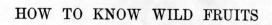

HOW TO KNOW WILD FRUITS

BLACK ALDER (*Ilex verticillata*)

HOW TO KNOW
WILD FRUITS

*A Guide to Plants when Not in Flower
by Means of Fruit and Leaf*

BY

MAUDE GRIDLEY
PETERSON

ILLUSTRATED BY

MARY ELISABETH HERBERT

WITH A NEW TABLE OF

CHANGES IN NOMENCLATURE BY

E. S. HARRAR

JAMES B. DUKE PROFESSOR OF WOOD SCIENCE

DUKE UNIVERSITY SCHOOL OF FORESTRY

DOVER PUBLICATIONS, INC.
NEW YORK

This Dover edition, first published in 1973, is an unabridged republication of the work originally published by The Macmillan Company in 1905. A new Table of Changes in Nomenclature by E. S. Harrar and a Publisher's Note have been added to this edition.

International Standard Book Number: 0-486-22943-2
Library of Congress Catalog Card Number: 72-95943

Manufactured in the United States of America
Dover Publications, Inc.
180 Varick Street
New York, N.Y. 10014

To the Mansfield and the Coventry hills,
from which its inspiration came, I would
reverently inscribe this book.

" On the motionless branches of some trees, autumn
berries hung like clusters of coral-beads, as in those
fabled orchards where the fruits were jewels."

—DICKENS, in *Martin Chuzzlewit*.

" A wonderful thing is a seed ;
  The one thing deathless forever —
Forever old and forever new,
Utterly faithful and utterly true —
  Fickle and faithless never."

PUBLISHER'S NOTE (1973)

In order to bring this work up to date and acknowledge the revisions in plant taxonomy that have been made over the years, the publisher has provided a Table of Changes in Nomenclature (pp. xlv ff.), prepared by Professor E. S. Harrar of the Duke University School of Forestry. Each species is listed, with its currently accepted common names, scientific name and family, in the order in which it appears in the main part of the book. The U.S. Forest Service's Checklist was used in tabulating the preferred common names and valid scientific names for all of the arborescent species treated. The preferred common and valid scientific names of the shrubs, vines and herbaceous species were drawn from the eighth edition of *Gray's Manual of Botany*, except where no common name for a species is listed. In such instances common names were drawn from *Standard Plant Names*, published by the American Joint Committee on Horticultural Nomenclature.

The reader should also note that a number of

changes have been made in family groupings, mainly the uniting of two or more formerly separate families. The Smilaceae (Smilax Family) and Convallariaceae (Lily-of-the-Valley Family) have been combined with the Liliaceae (Lily Family). Grossulariaceae (Gooseberry Family) has been united with Saxifragaceae (Saxifrage Family) and Vacciniaceae (Huckleberry Family) with Ericaceae (Heath Family). The previously separate Drupaceae (Plum Family) and Pomaceae (Apple Family) have both been combined with Rosaceae (Rose Family). The genus *Juniperus*, which was classified in the Pine Family, is now included in the Cupressaceae (Cedar Family), and the Holly Family, formerly called Ilicaceae, has been changed to Aquifoliaceae.

CONTENTS

CONTENTS

ILLUSTRATIONS

ix

INTRODUCTION

IF in country drive or ramble we happen upon an unknown flower, it is a comparatively easy matter, by means of the illustrations and the color guides of the modern field books of wild flowers, to identify it. The lack of similar reference books for identifying a plant by its fruit was forcibly brought to my notice during a drive in early autumn. Our journeyings led us along a wooded roadway where it was no longer the brilliance of the *flowers* which demanded our attention, but rather the attractive masses of *fruits*. There was one shrub bearing fruits of varying colors in different stages of development which was very attractive and which I did not know. I naturally wished to make its acquaintance.

Here the aforesaid field books failed to give their ready aid. Any system of analysis was of no avail, as the flower which preceded this special fruit was unobtainable. I was surprised at the meagerness of the descriptions of the fruits which I read, hoping to find my specimen among them. It was this difficulty of approach to the identification of my fruited plant, and the scarcity of material relating to this aspect of the plant's life, that suggested the present book.

I have attempted to deal with those plants only which bear attractively colored fruits. These fruits are the more noticeable ones; they do not, in most cases, develop until the blossoms have entirely disappeared; and they naturally fall into a class by themselves, being adapted for the same method of seed dispersal. The list will naturally include herbs, shrubs, and trees. A guide based on the kind and structure of the fruit will aid in determining the family to which a plant belongs, and under each family the species are grouped by colors. The illustrations will also aid in identifying specimens.

If the acquaintance of approximately two hundred plants of our northeastern section in their fruited stage is made more accessible; if added attention is attracted to the *result* of the work of the flower, making our knowledge of the cycle of the plant's life more complete, the work, fragmentary though it be, may have a place.

The order of arrangement of the Plant Families follows that of Engler and Prantl. The nomenclature and arrangement of species is essentially that of Britton and Brown. The additional name is the term used in Gray's sixth edition. In the classification of the Blackberries I have followed the general plan of L. H. Bailey, who has made a recent and careful study of them.

I am indebted to many a work of reference for aid: Gray's "Manual," Britton and Brown's "Illustrated Flora of the United States and Canada," Emerson's "Report of the Trees and Shrubs of Massachusetts," Card's "Bush Fruits," Bailey's "Evolution of our Native Fruits," Kerner and Oliver's "The Natural History of Plants," and others.

To the friends who have kindly furnished specimens I would extend my sincere gratitude. There have been many who, by the expression of a need for help such as the present book hopes to give, or by suggestion and encouragement, have strengthened my purpose to carry on the work to its fulfillment. I hold them all in grateful remembrance.

ADAPTATIONS OF FRUITS AND SEEDS FOR DISPERSAL AND PROTECTION

THE perfected fruit — how it suggests both the backward and the forward look : backward over the stages of growth which have produced it, forward to the stages of growth which are potential within it.

> " My heart is awed within me when I think
> Of the great miracle which still goes on
> In silence round me — the perpetual work
> Of thy creation, finished, yet renewed
> Forever."
> — BRYANT.

Miracle, indeed, it is that inclosed within the seed is the power to burst its bonds, to utilize its stored food material, to gather nutriment from earth and air, to grow and to produce again a seed capable of continuing the life of the species from which it has sprung.

For the development of this, however, certain conditions are necessary. The new plant must have room in earth and air for its best growth. If the seeds of a plant all started life in the immediate vicinity of the parent, and this were continued year after year, it is easy to see that there would soon be no room for new growths. The necessity of a means of dispersal seems to be fundamental to the fruit or seed. Examine fruit after fruit or seed after seed, and the varied and adequate structures for their dissemination are found to be most interesting.

The winged fruit of the maple, the tufted seed of the milkweed, and the plumed fruit of the clematis are a few of numerous examples of fruits fitted for dispersal by means of the wind.

Some fruits have mechanical devices which throw their seeds to a necessarily short distance. Have you not violated the command of the touch-me-not, just to see the parts of the pod curl up and throw out the seeds? This method of dispersal is not especially advantageous, and is confined to comparatively few plants. It is interesting to note that these, as a rule, grow in spots sheltered from the wind, where its agency would be unavailing in scattering their seeds.

Water is an agent in transporting a relatively small number of fruits. The cocoanut is admirably adapted for this means of dispersal, and the existence of the cocoanut palm on widely separated coral islands is interesting in this connection.

Doubtless your own experience with burdocks, agrimony, sticktights, and beggar ticks has been sufficiently emphatic to render unnecessary further mention of those fruits or seeds which, fastening themselves to men and animals, are carried by them hither and thither. Plants bearing such fruits naturally grow most profusely by the side of the road or footpath.

Animals and especially birds are instrumental in scattering seeds in still another way, using a part of the fruit for food and ejecting the seeds. It is this class of fruits with which this book is chiefly concerned.

That seed dispersal is accomplished by the means noted above has been a matter of dispute among botanists, but carefully conducted experiments have proved, without a doubt, that many birds eject the consumed seeds unharmed. Interesting accounts of these are given in Kerner and Oliver's " Natural History of Plants."

It has been found that while some birds grind up and destroy even the hardest coated seeds, others, like the ravens and jackdaws, destroy only the soft-coated seeds; while thrushes and blackbirds eject unharmed a large majority of the seeds eaten. The small seeds pass entirely through the intestinal tract while the larger ones are separated from the pulp in the crop, the pulp passing on into the gizzard and the seeds being thrown up.

Many plants whose seeds are scattered by birds grow along the fence rows. The significance of this is apparent, the seed being dropped by the bird as he rested upon the fence.

The parallel between the interdependence of the flower and the insect and that of the fruit and the bird is striking. The flower sets forth honey and sometimes a surplus of pollen for its guests, its color decorations are arranged most effectively, while often a subtle odor is a sign of welcome or repulse to wandering insects. The bee or other insect responds to these attractions and duly regales himself. In return for the hospitalities extended, he serves as the flower's messenger, bringing to the pistil of the flower-host pollen from a neighboring bloom, or bearing away with him freshly gathered pollen grains to deposit upon a near-by pistil. Cross fertilization, by means of which more vigorous seeds are produced, is thus accomplished.

Turning to the fruit, we find similar attractions offered to the birds. A pulp is usually developed for food, an odor is sometimes present, as in the case of the strawberry, grape, and pineapple, and the different color schemes are fascinating.

In what setting will the red fruit appear to greatest advantage ? In a green one, surely, and it is the plants whose fruit ripens while the surrounding foliage is still green, or whose foliage is evergreen, which usually bear red fruits. Amidst the brightly colored leafage of autumn, how effective are the blue and black drupes and berries ! Sometimes the dark-colored fruits are borne on red stems, producing a similar result. White fruits grow usually on plants which shed their leaves early, the white being brought into contrast with the bare branches, or, if low plants, with the floor covering of fallen leaves. The fruits are often massed in close heads, spikes, etc., rendering them still more conspicuous.

The bird recognizes the sign of his especial hostelry from afar, and comes to the feast spread for him. As we have seen, he renders his host a mutual service, depositing its offspring in various places, some of which will doubtless prove auspicious for the seed's development.

Bears are fond of berries and are said to scatter their seed. Mrs. Dana speaks of the berry of the wintergreen (*Gaultheria procumbens*) serving as food in winter for the hungry deer.

Until the seeds are ripe, many features serve to protect them from destruction by the agent afterward so useful in dispersing them. The fruit has a disagreeable taste, is sometimes even poisonous. It has no scent and is inconspicuous, being green, like the foliage. Some fruits, like the chestnut, walnut, etc., have an especial protection in the surrounding involucre.

Some ripe fruits have certain means of protection against foes that destroy the seed as well as the pulp. Mice are fond of rose hips and the contained seeds, but do not venture along the thorny way by which they are reached. Fallen cherries are eagerly eaten by centipedes, but hanging on their lengthened stalks they are comparatively safe from them.

Our knowledge is as yet not sufficiently detailed to state definitely that all of the fruits described in this work are used as food by birds or animals that scatter their seeds. Such fruits are included as seem adapted for transmission in this way. We do know that many of them are eaten by birds and know also the kind or kinds of birds using them. Dr. C. F. Hodge, in his book, "Nature Study and Life," has an interesting table of birds and their foods which includes a number of these fruits. Numerous investigations along this line are being made. Birds from various sections are sent to government experts, who, from an examination of the contents of their food tracts, are enabled to determine many of the foods eaten by them. Observation of living birds and the foods they choose is urged.

Full of meaning is the study of fruits and their adaptations, and includes many uninterpreted problems, making us feel, with Calderwood, " that the more we know, the more impressive becomes the unknown."

DEFINITIONS

THE *fruit* is the ripened ovary, its contents, and any other parts that are closely connected with it.

A *berry* is rather thin-skinned, and has its seeds loosely imbedded in soft pulpy or succulent material. An orange, a grape, a currant, are illustrations.

A *drupe* has for its distinguishing feature a stone inclosing the seed. The portion surrounding it may be fleshy, as in the peach; fibrous, as in the cocoanut; or leathery, as in the walnut.

A *pome* has its seeds and their cartilaginous or bony surrounding membranes inclosed in a fleshy mass, which is thickened calyx or some· times partly receptacle. Apple, pear, and quince are examples.

Aggregate fruits are masses of several carpels of the same flower which, when ripe, may or may not remain fast to the receptacle on which they are borne. Raspberry and blackberry are familiar examples.

Multiple fruits are compact masses of the ripened product of many flowers. Pineapple and mulberry are the usual illustrations.

Accessory fruits are simple fruits which have incorporated with them, as part of their mass, the developed surroundings or supports of the pistil. *Gaultheria* has its capsule surrounded by thickened fleshy calyx.

GUIDE TO PLANT FAMILIES
REPRESENTED

[In recent years a number of changes have been made in family groupings. The reader is advised to refer to the 1973 Publisher's Note (pages v–vi) for a full discussion of these changes.]

I. GYMNOSPERMÆ

II. ANGIOSPERMÆ

Pistil consists of closed ovary containing the ovules and becoming the fruit.

1. MONOCOTYLEDONES

Stem without central pith or annual layers, but with vascular bundles scattered through them.

Leaves are mostly parallel-veined.

Parts of flowers usually arranged in threes.

Embryo has but one cotyledon.

FRUIT A BERRY

2. DICOTYLEDONES

Stems have bark, wood, and pith.
Leaves are net-veined.
Parts of flowers in fours or fives.
Embryo has two or more cotyledons.

(*A*) Fruit a Berry

I. Calyx persistent.

PAGE OF
FAMILIES AND
SPECIES

(*a*) Berry crowned with shriveled remains of calyx.
Grossulariaceæ . . xxxvi

(*b*) Calyx teeth or top of tube crowning summit of fruit. Vaccinium, Oxycoccus, and Chiogenes in Vacciniaceæ . . . } xli, xlii

Fruit in clusters on stems from the axils of the leaves. Symphoricarpos and Lonicera in Caprifoliaceæ . . } xliii

(*c*) Calyx persistent at base of fruit and sometimes inclosing it.

Plumlike fruit containing 4 to 8 hard seeds. Calyx thickened.
Ebenaceæ xlii

Many seeds arranged around an axial placenta, which sometimes extends far into the cells. Solanaceæ xlii

Racemes. Berry 5- to 16-celled, one seed in each. Phytolaccaceæ . . xxxiv

II. Calyx absent.

(*a*) Several-seeded.

(*B*) FRUIT A DRUPE

Berrylike Drupe

I. Calyx persistent.

FAMILIES AND SPECIES

[In recent years a number of changes have been made in family groupings. The reader is advised to refer to the 1973 Publisher's Note (pages v–vi) for a full discussion of these changes.]

I. GYMNOSPERMÆ

PINACEÆ   **PINE FAMILY**

Blue

Juniperus communis L. . . . . Common Juniper.
Juniperus nana Willd. . . . . . Low Juniper.
Juniperus Virginiana L. . . . . Red Cedar.
Juniperus Sabina L. . . . . . Shrubby Red Cedar.

TAXACEÆ   **YEW FAMILY**

Red or Reddish Purple

Taxus minor (Michx.) Britton . American Yew.

II. ANGIOSPERMÆ

(a) MONOCOTYLEDONES

ARACEÆ   **ARUM FAMILY**

Red or Reddish Purple

Arisæma triphyllum (L.) Torr. Indian Turnip.
Arisæma dracontium (L.) Schott. Green Dragon.
Calla palustris L. . . . . . . Water-arum.

Green

Peltandra Virginica (L.) Kunth. Green Arrow-arum.

CONVALLARIACEÆ   **LILY–OF–THE–VALLEY FAMILY**

Red or Reddish Purple

Asparagus officinalis L. . . . Asparagus.
Vagnera racemosa (L.) Morong. Wild Spikenard.

Vagnera trifolia (L.) Morong.   Three-leaved Solomon's Seal.
Unifolium Canadense (Desf.)
    Greene . . . .   False Lily-of-the-Valley.
Disporum lanuginosum (Michx.)
    Nichols . . . .   Hairy Disporum.
Streptopus amplexifolius(L.)DC.   Clasping-leaved Twisted
               Stalk.
Streptopus roseus Michx. . .   Sessile-leaved Twisted Stalk.
Trillium sessile L. . . . . .   Sessile-flowered Wake-robin.
Trillium nivale Riddell . . .   Early Wake-robin.
Trillium grandiflorum (Michx.)
    Salisb. . . . .   Large-flowered Wake-robin.
Trillium erectum L. . . . .   Ill-scented Wake-robin.
Trillium cernuum L. . . . .   Nodding Wake-robin.
Trillium undulatum Willd. . .   Painted Wake-robin.

Black or Dark Purple

Clintonia umbellulata (Michx.)
    Torr. . . . .   White Clintonia.
Vagnera stellata (L.) Morong.   Star-flowered Solomon's Seal.
Polygonatum biflorum (Walt.)
    Ell. . . . .   Hairy Solomon's Seal.
Polygonatum commutatum (R.
    & S.) Dietr. . . . . .   Smooth Solomon's Seal.
Medeola Virginiana L. . . . .   Indian Cucumber Root.

Blue

Clintonia borealis (Ait.) Raf.   Yellow Clintonia.

SMILACEÆ   **SMILAX FAMILY**

Red or Reddish Purple

Smilax Walteri Pursh . . . .   Walter's Greenbrier.

Black or Dark Purple

Smilax herbacea L. . . . . .   Carrion Flower.
Smilax tamnifolia Michx. . .   Halberd-leaved Smilax.
Smilax glauca Walt. . . . .   Glaucous-leaved Greenbrier.

Smilax rotundifolia L.   . . .   Greenbrier.
Smilax hispida Muhl. . . . .   Hispid Greenbrier.
Smilax Pseudo-China L. . . .   Long-stalked Greenbrier.
Smilax Bona-nox L.   . . . .   Bristly Greenbrier.

(b) DICOTYLEDONES

MYRICACEÆ   **BAYBERRY FAMILY**
White
Myrica Carolinensis Mill . . .   Waxberry.

ULMACEÆ   **ELM FAMILY**
Black or Dark Purple
Celtis occidentalis L. . . . .   Sugarberry.

MORACEÆ   **MULBERRY FAMILY**
Black or Dark Purple
Morus rubra L. . . . . . .   Red Mulberry.
White
Morus alba L. . . . .   White Mulberry.

LORANTHACEÆ   **MISTLETOE FAMILY**
White
Razoumofskya pusilla (Peck)
    Kuntze . . . .   Small Mistletoe.
Phoradendron flavescens (Pursh)
    Nutt. . . . .   American Mistletoe.

PHYTOLACCACEÆ   **POKEWEED FAMILY**
Black or Dark Purple
Phytolacca decendra L. . . .   Poke.

MAGNOLIACEÆ   **MAGNOLIA FAMILY**
Red or Reddish Purple
Magnolia Virginiana L. . . .   Laurel Magnolia.
Magnolia acuminata L. . . .   Cucumber Tree.

ANONACEÆ   **CUSTARD-APPLE FAMILY**

Yellow

Asimina triloba (L.) Dunal.   .   North American Papaw.

RANUNCULACEÆ   **CROWFOOT FAMILY**

Red or Reddish Purple

Hydrastis Canadensis L. .   .   .   Orange Root.
Actæa rubra (Ait.) Willd. .   .   Red Baneberry.

White

Actæa alba (L.) Mill.   .   .   .   .   White Baneberry.

BERBERIDACEÆ   **BARBERRY FAMILY**

Red or Reddish Purple

Berberis vulgaris L.   .   .   .   .   Common Barberry.

Blue

Caulophyllum thalictroides (L.)
    Michx. .   .   .   .   Blue Cohosh.

Yellow

Podophyllum peltatum L. .   .   .   May Apple.

MENISPERMACEÆ   **MOONSEED FAMILY**

Black or Dark Purple

Menispermum Canadense L.   .   Canada Moonseed.

LAURACEÆ   **LAUREL FAMILY**

Red or Reddish Purple

Benzoin Benzoin (L.) Coulter   .   Spice Bush.

Blue

Sassafras Sassafras (L.) Karst. .   Sassafras.

GROSSULARIACEÆ   **GOOSEBERRY FAMILY**

Red or Reddish Purple

Ribes oxyacanthoides L. . . .   Hawthorn or Northern Gooseberry.
Ribes rotundifolium Michx.   .   Eastern Wild Gooseberry.
Ribes lacustre (Pers.) Poir. . .   Swamp Gooseberry.
Ribes prostratum L'Her. . . .   Fetid Currant.
Ribes rubrum L. . . . . . .   Red Currant.

Black or Dark Purple

Ribes Cynosbati L. . . . . .   Wild Gooseberry.
Ribes floridum L'Her. . . . .   Wild Black Currant.

ROSACEÆ   **ROSE FAMILY**
Red or Reddish Purple

Rubus odoratus L. . . . . .   Purple-flowering Raspberry.
Rubus Chamæmorus L. . . .   Cloudberry.
Rubus strigosus Michx. . . .   Wild Red Raspberry.
Rubus neglectus Peck . . . .   Purple Wild Raspberry.
Rubus Americanus (Pers.) Brit-
ton . . . .   Dwarf Raspberry.
Fragaria Virginiana Duchesne .   Virginia or Scarlet Strawberry.
Fragaria Canadensis Michx. .   Northern Wild Strawberry.
Fragaria vesca L. . . . . .   European Wood Strawberry.
Fragaria Americana (Porter)
Britton . . . .   American Wood Strawberry.
Rosa blanda Ait. . . . . . .   Smooth or Meadow Rose.
Rosa Carolina L. . . . . . .   Swamp Rose.
Rosa humilis Marsh. . . . .   Low or Pasture Rose.
Rosa nitida Willd. . . . . .   Northeastern Rose.
Rosa canina L. . . . . . .   Dog Rose.
Rosa rubiginosa L. . . . . .   Sweetbrier.

Black or Dark Purple

Rubus occidentalis L.   . . .   Black Raspberry.
Rubus villosus Ait.   . . . .   Low Blackberry.

Rubus hispidus L.	Running Swamp Blackberry.
Rubus cuneifolius Pursh . . .	Sand Blackberry.
Rubus nigrobaccus	High Bush Blackberry.
Rubus nigrobaccus, var. sativus	Short Cluster Blackberry.
Rubus Allegheniensis Porter .	Mountain Blackberry.
Rubus argutus Link.	Leafy Cluster Blackberry.
Rubus Canadensis L.	Thornless Blackberry.

POMACEÆ   **APPLE FAMILY**

Red or Reddish Purple

Sorbus Americana Marsh. . .	American Mountain Ash.
Sorbus sambucifolia (C. & S.) Roem.	Western Mountain Ash.
Aronia arbutifolia (L.) Ell. . .	Red Chokeberry.
Amelanchier Botryapium (L. f.) DC.	Shad Bush.
Amelanchier Canadensis (L.) Medic.	Juneberry.
Cratægus Crus-Galli L. . . .	Cockspur Thorn.
Cratægus punctata Jacq. . .	Large-fruited Thorn.
Cratægus coccinea L.	Scarlet Thorn.
Cratægus macracantha Lodd. .	Long-spined Thorn.
Cratægus mollis (T. & G.) Scheele	Red-fruited Thorn.
Cratægus tomentosa L. . . .	Pear Thorn.

Black or Dark Purple

Aronia nigra (Willd.) Britton .	Black Chokeberry.
Amelanchier oligocarpa (Michx.) Roem.	Oblong-fruited Juneberry.

Yellow

Cratægus uniflora Muench. . .	Dwarf Thorn.

Green

Pyrus communis L.	Choke pear.
Malus coronaria (L.) Mill. . .	American Crab Apple.
Malus angustifolia (Ait.) Michx.	Narrow-leaved Crab Apple.

DRUPACEÆ   **PLUM FAMILY**
Red or Reddish Purple

Prunus Americana Marsh. . . Wild Yellow or Red Plum.
Prunus nigra Ait. . . . . . Canada Plum.
Prunus maritima Wang. . . Beach Plum.
Prunus Pennsylvanica L. f. . . Wild Red Cherry.
Prunus Virginiana L. . , . . Chokecherry.

Black or Dark Purple

Prunus Allegheniensis Porter . Porter's Plum.
Prunus spinosa L. . . . . . Sloe.
Prunus pumila L. . . . . . Dwarf Cherry.
Prunus serotina Ehrh. . . . . Wild Black Cherry.

EMPETRACEÆ   **CROWBERRY FAMILY**
Black or Dark Purple

Empetrum nigrum L. . . . . Black Crowberry.

ANACARDIACEÆ   **SUMAC FAMILY**
Red or Reddish Purple

Rhus copallina L. . . . . . Dwarf Sumac.
Rhus hirta (L.) Sudw. . . . . Staghorn Sumac.
Rhus glabra L. . . . . . . Smooth Sumac.
Rhus aromatica Ait. . . . . Fragrant or Sweet-scented Sumac.

White

Rhus Vernix L. . . . . . . Poison Sumac.
Rhus radicans L. . . . . . . Poison, Climbing, or Three-leaved Ivy.

ILICACEÆ   **HOLLY FAMILY**
Red or Reddish Purple

Ilex opaca Ait. . . . . . . American Holly.
Ilex monticola A. Gray . . . Large-leaved Holly.
Ilex verticillata (L.) A. Gray . Black Alder.
Ilex lævigata (Pursh) A. Gray . Smooth Winter Berry.
Ilicioides mucronata (L.) Britton Wild or Mountain Holly.

Black or Dark Purple

Ilex glabra (L.) A. Gray . . .   Inkberry.

CELASTRACEÆ   **STAFF-TREE FAMILY**

Red or Reddish Purple

Euonymus Americanus L. . .   Strawberry Bush.
Euonymus obovatus Nutt. . .   Running Strawberry Bush.
Euonymus atropurpureus Jacq.. .   Burning Bush.
Celastrus scandens L. . . . .   Shrubby or Climbing Bitter-
                                    sweet.

RHAMNACEÆ   **BUCKTHORN FAMILY**

Black or Dark Purple

Rhamnus cathartica L. . . .   Buckthorn.
Rhamnus lanceolata Pursh . .   Lance-leaved Buckthorn.
Rhamnus alnifolia L'Her.. . .   Alder-leaved Buckthorn.

VITACEÆ   **GRAPE FAMILY**

Black or Dark Purple

Vitis Labrusca L. . . . . .   Northern Fox Grape.
Vitis æstivalis Michx. . . . .   Summer Grape.
Vitis bicolor LeConte . . . .   Blue Grape.
Vitis vulpina L. . . . . . .   Riverside or Sweet-scented
                                    Grape.
Vitis cordifolia Michx. . . . .   Frost or Chicken Grape.
Parthenocissus quinquefolia (L.)
    Planch.. . . . .   Virginia Creeper.

THYMELEACEÆ   **MEZEREON FAMILY**

Red or Reddish Purple

Dirca palustris L. . . . . .   Leatherwood.

ELÆAGNACEÆ   **OLEASTER FAMILY**

Red or Reddish Purple

Lepargyræa Canadensis   (L.)
    Greene . . . .   Canadian Buffalo Berry.

ARALIACEÆ   **GINSENG FAMILY**
Red or Reddish Purple
Panax quinquefolium L.   .   .   .   Ginseng.

Black or Dark Purple
Aralia spinosa L. .   .   .   .   .   .   Hercules' Club.
Aralia racemosa L.   .   .   .   .   American Spikenard.
Aralia nudicaulis L. .   .   .   .   .   Wild or Virginian Sarsapa-
                                        rilla.
Aralia hispida Vent.   .   .   .   .   Bristly Sarsaparilla.

Yellow
Panax trifolium L.   .   .   .   .   .   Dwarf Ginseng.

CORNACEÆ   **DOGWOOD FAMILY**
Red or Reddish Purple
Cornus Canadensis L.   .   .   .   .   Low or Dwarf Cornel.
Cornus florida L. .   .   .   .   .   .   Flowering Dogwood.

Black or Dark Purple
Nyssa sylvatica Marsh. .   .   .   .   Tupelo.

Blue
Cornus circinata L'Her.   .   .   .   Round-leaved Cornel or Dog-
                                        wood.
Cornus Amonum Mill. .   .   .   .   Silky Cornel.
Cornus alternifolia L. f.   .   .   .   Alternate-leaved Cornel or
                                        Dogwood.

White
Cornus stolonifera Michx. .   .   .   Red-osier Cornel or Dog-
                                        wood.
Cornus candidissima Marsh. .   .   Panicled Cornel or Dogwood.

ERICACEÆ   **HEATH FAMILY**
Red or Reddish Purple
Gaultheria procumbens L.   .   .   Spring or Creeping Winter-
                                        green.

Arctostaphylos   Uva-Ursi   (L.)
    Spreng. . . . .   Red Bearberry.

Black or Dark Purple

Mairania alpina (L.) Desv. . .   Alpine or Black Bearberry.

VACCINIACEÆ   **HUCKLEBERRY FAMILY**

Red or Reddish Purple

Vaccinium Vitis Idæa L. . . .   Mountain Cranberry.
Oxycoccus oxycoccus (L.) MacM.   Small or European Cran-
                                        berry.
Oxycoccus   macrocarpus   (Ait.)
    Pers. . . . .   Large or American Cran-
                                        berry.

Black or Dark Purple

Gaylussacia resinosa (Ait.) T.
    & G. . . . .   Black or High-bush Huckle-
                                        berry.
Gaylussacia dumosa (Andr.) T.
    & G. . . . .   Dwarf or Bush Huckleberry.
Vaccinium atrococcum (A. Gray)
    Heller . . . .   Black Blueberry.
Vaccinium nigrum (Wood) Brit-
    ton . . . .   Low Black Blueberry.

Blue

Gaylussacia frondosa (L.) T. &
    G. . . . .   Blue Tangle.
Gaylussacia brachycera (Michx.)
    A. Gray . . . .   Box Huckleberry.
Vaccinium uliginosum L. . . .   Great Bilberry.
Vaccinium cæspitosum Michx.   Dwarf Bilberry.
Vaccinium corymbosum L. . .   High-bush or Tall Blueberry.
Vaccinium Pennsylvanicum Lam.   Dwarf Blueberry.
Vaccinium Canadense Richards.   Canada Blueberry.
Vaccinium vacillans Kalm. . .   Low Blueberry.

Yellow

Vaccinium stamineum L. . . . Deerberry.

White

Chiogenes hispidula (L.) T. & G. Creeping Snowberry.

EBENACEÆ   **EBONY FAMILY**

Yellow

Diospyros Virginiana L. . . . Persimmon.

OLEACEÆ   **OLIVE FAMILY**

Black or Dark Purple

Chionanthus Virginica L. . . Fringe Tree.
Ligustrum vulgare L. . . . . Privet.

SOLANACEÆ   **POTATO FAMILY**

Red or Reddish Purple

Physalis Philadelphica Lam. . Philadelphia Ground Cherry.
Solanum Dulcamara L. . . . Nightshade.
Lycium vulgare (Ait. f.) Dunal. Matrimony Vine.

Black or Dark Purple

Solanum nigrum L. . . . . Black or Garden Nightshade.

Yellow

Physalis pubescens L. . . . . Low Hairy Ground Cherry.
Physalis angulata L. . . . . Cut-leaved Ground Cherry.
Physalis heterophylla Nees. . . Clammy Ground Cherry.
Solanum Carolinense L. . . . Horse Nettle.

RUBIACEÆ   **MADDER FAMILY**

Red or Reddish Purple

Mitchella repens L. . . . . . Partridge Berry.

CAPRIFOLIACEÆ   **HONEYSUCKLE FAMILY**

Red or Reddish Purple

Sambucus pubens Michx. . . . Red-berried Elder.
Viburnum alnifolium Marsh. . Hobble Bush.

Viburnum Opulus L. . . . . Cranberry Tree.
Viburnum pauciflorum Pylaie . Few-flowered Cranberry Tree.
Triosteum perfoliatum L. . . Feverwort.
Symphoricarpos Symphoricarpos
   (L.) MacM. . . . . . . Coral Berry.
Lonicera Caprifolium L. . . . Italian or Perfoliate Honey-
                      suckle.
Lonicera hirsuta Eaton . . . Hairy Honeysuckle.
Lonicera dioica L. . . . . . Smooth-leaved or Glaucous
                      Honeysuckle.
Lonicera sempervirens L. . . Trumpet or Coral Honey-
                      suckle.
Lonicera oblongifolia (Goldie)
   Hook. . . . . Swamp Fly Honeysuckle.
Lonicera ciliata Muhl. . . . . American Fly Honeysuckle.

Black or Dark Purple

Sambucus Canadensis L. . . . American Elder.
Viburnum acerifolium L. . . . Maple-leaved Arrowwood.
Viburnum pubescens (Ait.) Pursh Downy-leaved Arrowwood.
Viburnum cassinoides L. . . . Withe-rod.
Viburnum nudum L. . . . . Larger Withe-rod.
Viburnum Lentago L. . . . . Nannyberry.
Viburnum prunifolium L. . . Black Haw.

Blue

Viburnum dentatum L. . . . Arrowwood.
Viburnum molle Michx. . . . Soft-leaved Arrowwood.
Lonicera cœrulea L. . . . . Blue or Mountain Fly Honey-
                      suckle.

White

Symphoricarpos racemosus
   Michx. . . . . Snowberry.
Symphoricarpos pauciflorus (Rob-
   bins) Britton . . . . . . Low Snowberry.

TABLE OF CHANGES IN NOMENCLATURE

Peterson Nomenclature	Current Nomenclature	
AMERICAN YEW *Taxus minor, Taxus Canadensis* Yew Family	AMERICAN YEW *Taxus canadensis* Yew Family	3
INDIAN TURNIP, JACK-IN-THE-PULPIT *Arisaema triphyllum* Arum Family	NO CHANGE	5
GREEN DRAGON, DRAGON ROOT *Arisaema Dracontium* Arum Family	DRAGON-ROOT *Arisaeum dracontium* Arum Family	8
WATER ARUM *Calla palustris* Arum Family	WATER-ARUM, WILD CALLA *Calla palustris* Arum Family	9
ASPARAGUS *Asparagus officinalis* Lily-of-the-Valley Family	ASPARAGUS *Asparagus officinalis* Lily Family	10
WILD SPIKENARD *Vagnera racemosa, Smilacina racemosa* Lily-of-the-Valley Family	FALSE SPIKENARD *Smilacina racemosa* Lily Family	11
THREE-LEAVED SOLOMON'S SEAL *Vagnera trifolia, Smilacina trifolia* Lily-of-the-Valley Family	THREE-LEAVED SOLOMON'S SEAL *Smilacina trifolia* Lily Family	13

Peterson Nomenclature	Current Nomenclature	
FALSE LILY-OF-THE-VALLEY, TWO-LEAVED SOLOMON'S SEAL	FALSE LILY-OF-THE-VALLEY	14
Unifolium Canadense, Maianthemum Canadense	*Maianthemum canadense*	
Lily-of-the-Valley Family	Lily Family	
HAIRY DISPORUM	YELLOW MANDARIN	16
Disporum lanuginosum	*Disporum lanuginosum*	
Lily-of-the-Valley Family	Lily Family	
CLASPING-LEAVED TWISTED STALK	WHITE MANDARIN	16
Streptopus amplexifolius	*Streptopus amplexifolius*	
Lily-of-the-Valley Family	Lily Family	
SESSILE-LEAVED TWISTED STALK	ROSE MANDARIN	18
Streptopus roseus	*Streptopus roseus*	
Lily-of-the-Valley Family	Lily Family	
SESSILE-FLOWERED WAKE-ROBIN	TOADSHADE	19
Trillium sessile	*Trillium sessile*	
Lily-of-the-Valley Family	Lily Family	
EARLY WAKE-ROBIN	DWARF WHITE TRILLIUM	19
Trillium nivale	*Trillium nivale*	
Lily-of-the-Valley Family	Lily Family	
LARGE-FLOWERED WAKE-ROBIN	LARGE-FLOWERED TRILLIUM	20
Trillium grandiflorum	*Trillium grandiflorum*	
Lily-of-the-Valley Family	Lily Family	
ILL-SCENTED WAKE-ROBIN, BIRTHROOT	PURPLE TRILLIUM, SQUAWROOT	21
Trillium erectum	*Trillium erectum*	
Lily-of-the-Valley Family	Lily Family	
NODDING WAKE-ROBIN	NODDING TRILLIUM	23
Trillium cernuum	*Trillium cernuum*	
Lily-of-the-Valley Family	Lily Family	

Peterson Nomenclature	Current Nomenclature	
PAINTED WAKE-ROBIN	PAINTED TRILLIUM	24
Trillium undulatum, Trillium erythrocarpum	*Trillium undulatum*	
Lily-of-the-Valley Family	Lily Family	
WALTER'S GREENBRIER	WALTER'S GREENBRIER	25
Smilax Walteri	*Smilax walteri*	
Smilax Family	Lily Family	
LAUREL MAGNOLIA	SWEETBAY	26
Magnolia Virginiana, Magnolia glauca	*Magnolia virginiana*	
Magnolia Family	Magnolia Family	
CUCUMBER TREE, MOUNTAIN MAGNOLIA	CUCUMBERTREE	29
Magnolia acuminata	*Magnolia acuminata*	
Magnolia Family	Magnolia Family	
YELLOW PUCCOON, YELLOW ROOT, YELLOW INDIAN PAINT, ORANGE ROOT, GOLDEN SEAL	ORANGEROOT	30
Hydrastis Canadensis	*Hydrastis canadensis*	
Crowfoot Family	Crowfoot Family	
RED BANEBERRY	RED BANEBERRY	33
Actaea rubra, Actaea spicata, Var. *rubra*	*Actaea rubra*	
Crowfoot Family	Crowfoot Family	
COMMON BARBERRY	NO CHANGE	34
Berberis vulgaris		
Barberry Family		
WILD ALLSPICE, BENJAMIN BUSH, FEVER BUSH, SPICE BUSH	SPICEBUSH	38
Benzoin Benzoin, Lindera Benzoin	*Lindera benzoin*	
Laurel Family	Laurel Family	

Peterson Nomenclature	Current Nomenclature	
HAWTHORN OR NORTHERN GOOSEBERRY	NORTHERN GOOSEBERRY	40
Ribes oxyacanthoides	*Ribes oxyacanthoides*	
Gooseberry Family	Saxifrage Family	
EASTERN WILD GOOSEBERRY	EASTERN WILD GOOSEBERRY	42
Ribes rotundifolium	*Ribes rotundifolium*	
Gooseberry Family	Saxifrage Family	
SWAMP GOOSEBERRY	BRISTLY BLACK CURRANT	43
Ribes lacustre	*Ribes lacustre*	
Gooseberry Family	Saxifrage Family	
FETID CURRANT, MOUNTAIN CURRANT, PROSTRATE CURRANT	SKUNK CURRANT	44
Ribes prostratum	*Ribes glandulosum*	
Gooseberry Family	Saxifrage Family	
RED CURRANT	RED CURRANT	45
Ribes rubrum, Ribes rubrum, Var. subglandulosum	*Ribes sativum*	
Gooseberry Family	Saxifrage Family	
PURPLE-FLOWERING RASPBERRY	NO CHANGE	45
Rubus odoratus		
Rose Family		
CLOUDBERRY, BAKED-APPLE BERRY, MOUNTAIN RASPBERRY	CLOUDBERRY	47
Rubus Chamaemorus	*Rubus chamaemorus*	
Rose Family	Rose Family	
WILD RED RASPBERRY	WILD RED RASPBERRY	49
Rubus strigosus	*Rubus idaeus var. strigosus*	
Rose Family	Rose Family	
DWARF RASPBERRY	DWARF RASPBERRY	50
Rubus Americanus, Rubus triflorus	*Rubus pubescens*	
Rose Family	Rose Family	

Peterson Nomenclature	Current Nomenclature	
VIRGINIA OR SCARLET STRAWBERRY *Fragaria Virginiana* Rose Family	VIRGINIA STRAWBERRY *Fragaria virginiana* Rose Family	52
EUROPEAN WOOD STRAWBERRY *Fragaria vesca* Rose Family	EUROPEAN STRAWBERRY *Fragaria vesca* Rose Family	54
SMOOTH OR MEADOW ROSE *Rosa blanda* Rose Family	MEADOW ROSE *Rosa blanda* Rose Family	57
SWAMP ROSE *Rosa Carolina* Rose Family	SWAMP ROSE *Rosa carolina* Rose Family	58
LOW OR PASTURE ROSE *Rosa humilis* Rose Family	NOW COMBINED WITH *Rosa carolina*	59
NORTHEASTERN ROSE *Rosa nitida* Rose Family	SHINING ROSE *Rosa nitida* Rose Family	61
DOG ROSE, CANKER ROSE *Rosa canina* Rose Family	NO CHANGE	61
SWEETBRIER, EGLANTINE *Rosa rubiginosa* Rose Family	SWEETBRIER *Rosa eglanteria* Rose Family	62
AMERICAN MOUNTAIN ASH *Sorbus Americana, Pyrus Americana* Apple Family	AMERICAN MOUNTAIN-ASH *Sorbus americana* Rose Family	64

Peterson Nomenclature	Current Nomenclature	
RED CHOKEBERRY, DOGBERRY *Aronia arbutifolia, Pyrus arbutifolia* Apple Family	RED CHOKEBERRY *Aronia arbutifolia* Rose Family	66
SERVICE BERRY, JUNEBERRY, MAY CHERRY *Amelanchier Canadensis* Apple Family	DOWNY SERVICEBERRY *Amelanchier arborea* Rose Family	71
COCKSPUR THORN *Crataegus Crus-Galli* Apple Family	COCKSPUR HAWTHORN *Crataegus crus-galli* Rose Family	73
LARGE-FRUITED THORN, DOTTED-FRUITED THORN *Crataegus punctata* Apple Family	DOTTED HAWTHORN *Crataegus punctata* Rose Family	74
SCARLET THORN *Crataegus coccinea* Apple Family	SCARLET HAWTHORN *Crataegus pedicellata* Rose Family	75
PEAR THORN *Crataegus tomentosa* Apple Family	PEAR HAWTHORN *Crataegus calpodendron* Rose Family	78
WILD YELLOW OR RED PLUM *Prunus Americana* Plum Family	AMERICAN PLUM *Prunus americana* Rose Family	79
CANADA PLUM, HORSE PLUM *Prunus nigra* Plum Family	CANADA PLUM *Prunus nigra* Rose Family	81
BEACH PLUM *Prunus maritima* Plum Family	BEACH PLUM *Prunus maritima* Rose Family	82

Peterson Nomenclature	Current Nomenclature	
WILD RED CHERRY, BIRD CHERRY, PIN OR PIGEON CHERRY	PIN CHERRY	83
Prunus Pennsylvanica	*Prunus pensylvanica*	
Plum Family	Rose Family	
CHOKE CHERRY	CHOKECHERRY	86
Prunus Virginiana	*Prunus virginiana*	
Plum Family	Rose Family	
DWARF SUMAC	NO CHANGE	88
Rhus copallina		
Sumac Family		
STAGHORN SUMAC	STAGHORN SUMAC	89
Rhus hirta, Rhus typhina	*Rhus typhina*	
Sumac Family	Sumac Family	
SMOOTH SUMAC	NO CHANGE	92
Rhus glabra		
Sumac Family		
FRAGRANT OR SWEET-SCENTED SUMAC	FRAGRANT SUMAC	93
Rhus aromatica, Rhus Canadensis	*Rhus aromatica*	
Sumac Family	Sumac Family	
AMERICAN HOLLY	NO CHANGE	94
Ilex opaca		
Holly Family		
BLACK ALDER, VIRGINIA WINTER BERRY	COMMON WINTERBERRY	97
Ilex verticillata	*Ilex verticillata*	
Holly Family	Holly Family	
SMOOTH WINTER BERRY	SMOOTH WINTERBERRY	99
Ilex laevigata	*Ilex laevigata*	
Holly Family	Holly Family	

lii **TABLE OF CHANGES IN NOMENCLATURE**

Peterson Nomenclature	Current Nomenclature	
WILD OR MOUNTAIN HOLLY *Ilicioides mucronata,* *Nemopanthes fascicularis* Holly Family	MOUNTAIN-HOLLY *Nemopanthus mucronata* Holly Family	100
STRAWBERRY BUSH *Euonymus Americanus* Staff-tree Family	STRAWBERRY-BUSH *Euonymus americanus* Staff-tree Family	100
BURNING BUSH, WAHOO, SPINDLE TREE *Euonymus atropurpureus* Staff-tree Family	EASTERN WAHOO *Euonymus atropurpureus* Staff-tree Family	101
WAXWORK, SHRUBBY OR CLIMBING BITTERSWEET *Celastrus scandens* Staff-tree Family	AMERICAN BITTERSWEET *Celastris scandens* Staff-tree Family	103
LEATHERWOOD, MOOSEWOOD *Dirca palustris* Mezereon Family	LEATHERWOOD, MOOSEWOOD *Dirca palustris* Mezereum Family	105
CANADIAN BUFFALO BERRY *Lepargyraea Canadensis,* *Shepherdia Canadensis* Oleaster Family	SOAPBERRY *Shepherdia canadensis* Oleaster Family	106
GINSENG *Panax quinquefolium, Aralia* *quinquefolia* Ginseng Family	GINSENG *Panax quinquifolius* Ginseng Family	107
LOW OR DWARF CORNEL, BUNCHBERRY *Cornus Canadensis* Dogwood Family	BUNCHBERRY *Cornus canadensis* Dogwood Family	108
FLOWERING DOGWOOD *Cornus florida* Dogwood Family	NO CHANGE	110

Peterson Nomenclature	Current Nomenclature	
SPRING OR CREEPING WINTER-GREEN, CHECKERBERRY, BOXBERRY, TEABERRY	TEABERRY	114
Gaultheria procumbens	*Gaultheria procumbens*	
Heath Family	Heath Family	
RED BEARBERRY	COMMON BEARBERRY, KINNIKINICK	116
Arctostaphylos Uva-Ursi	*Arctostaphylos uva-ursi*	
Heath Family	Heath Family	
COWBERRY, MOUNTAIN CRANBERRY, FOXBERRY	COWBERRY	118
Vaccinium Vitis-Idaea	*Vaccinium vitis-idaea*	
Huckleberry Family	Heath Family	
SMALL OR EUROPEAN CRANBERRY	SMALL CRANBERRY	119
Oxycoccus Oxycoccus, Vaccinium Oxycoccus	*Vaccinum oxycoccos*	
Huckleberry Family	Heath Family	
LARGE OR AMERICAN CRANBERRY	LARGE CRANBERRY	121
Oxycoccus macrocarpus, Vaccinium macrocarpon	*Vaccinium macrocarpum*	
Huckleberry Family	Heath Family	
PHILADELPHIA GROUND CHERRY	PHILADELPHIA GROUND CHERRY	124
Physalis Philadelphica	*Physalis philadelphica*	
Potato Family	Nightshade Family	
NIGHTSHADE, BITTERSWEET	NIGHTSHADE	125
Solanum Dulcamara	*Solanum dulcamara*	
Potato Family	Nightshade Family	
MATRIMONY VINE	MATRIMONY VINE	129
Lycium vulgare	*Lycium halimifolium*	
Potato Family	Nightshade Family	

Peterson Nomenclature	Current Nomenclature	
PARTRIDGE BERRY *Mitchella repens* Madder Family	PARTRIDGE-BERRY *Mitchella repens* Madder Family	129
RED-BERRIED ELDER *Sambucus pubens, Sambucus racemosa* Honeysuckle Family	RED-BERRIED ELDER *Sambucus pubens* Honeysuckle Family	133
HOBBLE BUSH, WAYFARING TREE *Viburnum alnifolium, Viburnum lantanoides* Honeysuckle Family	HOBBLEBUSH, WITCH-HOBBLE *Viburnum alnifolium* Honeysuckle Family	135
CRANBERRY TREE, GUELDER ROSE *Viburnum Opulus* Honeysuckle Family	GUELDER-ROSE, HIGHBUSH CRANBERRY *Viburnum opulus* Honeysuckle Family	139
FEW-FLOWERED CRANBERRY TREE *Viburnum pauciflorum* Honeysuckle Family	SQUASHBUSH *Viburnum edule* Honeysuckle Family	140
TINKER'S WEED, WILD OR WOOD IPECAC, WILD COFFEE, HORSE GINSENG, FEVERWORT *Triosteum perfoliatum* Honeysuckle Family	TINKER'S-WEED *Triosteum perfoliatum* Honeysuckle Family	141
INDIAN CURRANT, CORAL BERRY *Symphoricarpos Symphoricarpos, Symphoricarpos vulgaris* Honeysuckle Family	CORALBERRY *Symphoricarpos orbiculatus* Honeysuckle Family	142
AMERICAN WOODBINE, ITALIAN OR PERFOLIATE HONEYSUCKLE *Lonicera Caprifolium, Lonicera grata* Honeysuckle Family	ITALIAN WOODBINE *Lonicera caprifolium* Honeysuckle Family	144

Peterson Nomenclature	Current Nomenclature	
HAIRY HONEYSUCKLE *Lonicera hirsuta* Honeysuckle Family	NO CHANGE	145
SMOOTH-LEAVED OR GLAUCOUS HONEYSUCKLE *Lonicera dioica, Lonicera glauca* Honeysuckle Family	LIMBER HONEYSUCKLE *Lonicera dioica* Honeysuckle Family	145
TRUMPET HONEYSUCKLE, CORAL HONEYSUCKLE *Lonicera sempervirens* Honeysuckle Family	TRUMPET HONEYSUCKLE *Lonicera sempervirens* Honeysuckle Family	148
SWAMP FLY HONEYSUCKLE *Lonicera oblongifolia* Honeysuckle Family	NO CHANGE	150
AMERICAN FLY HONEYSUCKLE *Lonicera ciliata* Honeysuckle Family	AMERICAN FLY HONEYSUCKLE *Lonicera canadensis* Honeysuckle Family	151
WHITE CLINTONIA *Clintonia umbellulata* Lily-of-the-Valley Family	SPECKLED WOOD-LILY *Clintonia umbellulata* Lily Family	155
STAR-FLOWERED SOLOMON'S SEAL *Vagnera stellata, Smilacina stellata* Lily-of-the-Valley Family	STAR-FLOWERED SOLOMON'S SEAL *Smilacina stellata* Lily Family	156
HAIRY SOLOMON'S SEAL *Polygonatum biflorum* Lily-of-the-Valley Family	HAIRY SOLOMON'S SEAL *Polygonatum canaliculatum* Lily Family	157
SMOOTH SOLOMON'S SEAL *Polygonatum commutatum, Polygonatum giganteum* Lily-of-the-Valley Family	SMOOTH SOLOMON'S SEAL *Polygonatum commutatum* Lily Family	159

Peterson Nomenclature	Current Nomenclature	
INDIAN CUCUMBER ROOT *Medeola Virginiana* Lily-of-the-Valley Family	INDIAN CUCUMBER ROOT *Medeola virginiana* Lily Family	161
CARRION FLOWER *Smilax herbacea* Smilax Family	CARRION FLOWER *Smilax herbacea* Lily Family	163
HALBERD-LEAVED SMILAX *Smilax tamnifolia* Smilax Family	HALBERD-LEAVED SMILAX *Smilax tamnifolia* Lily Family	167

GLAUCOUS-LEAVED GREENBRIER,
  FALSE SARSAPARILLA
Smilax glauca
Smilax Family
	SAWBRIER, WILD
	  SARSAPARILLA
	Smilax glauca
	Lily Family
		167

GREENBRIER, CATBRIER, HORSEBRIER *Smilax rotundifolia* Smilax Family	COMMON GREENBRIER *Smilax rotundifolia* Lily Family	168
HISPID GREENBRIER *Smilax hispida* Smilax Family	HISPID GREENBRIER *Smilax hispida* Lily Family	169
LONG-STALKED GREENBRIER *Smilax Pseudo-China* Smilax Family	LONG-STALKED GREENBRIER *Smilax pseudo-china* Lily Family	170

BRISTLY GREENBRIER,
  STRETCH BERRY
Smilax Bona-nox
Smilax Family
	BRISTLY GREENBRIER,
	  STRETCH BERRY
	Smilax bona-nox
	Lily Family
		171

HACKBERRY, SUGARBERRY *Celtis occidentalis* Elm Family	HACKBERRY *Celtis occidentalis* Elm Family	172

Peterson Nomenclature	Current Nomenclature	
RED MULBERRY *Morus rubra* Mulberry Family	NO CHANGE	173
POKE, SCOKE, GARGET, PIGEON BERRY *Phytolacca decendra* Pokeweed Family	POKE *Phytolacca americana* Pokeweed Family	176
CANADA MOONSEED *Menispermum Canadense* Moonseed Family	CANADA MOONSEED *Menispermum canadense* Moonseed Family	178
WILD GOOSEBERRY *Ribes Cynosbati* Gooseberry Family	WILD GOOSEBERRY *Ribes cynosbati* Saxifrage Family	179
WILD BLACK CURRANT *Ribes floridum* Gooseberry Family	WILD BLACK CURRANT *Ribes americanum* Saxifrage Family	180
BLACK RASPBERRY, THIMBLE BERRY *Rubus occidentalis* Rose Family	NO CHANGE	183
LOW RUNNING BLACKBERRY *Rubus villosus, Rubus Canadensis* Rose Family	LOW RUNNING BLACKBERRY *Rubus flagellaris* Rose Family	185
RUNNING SWAMP BLACKBERRY *Rubus hispidus* Rose Family	NO CHANGE	187
SAND BLACKBERRY *Rubus cuneifolius* Rose Family	NO CHANGE	188

Peterson Nomenclature	Current Nomenclature	
COMMON OR HIGH-BUSH BLACKBERRY *Rubus nigrobaccus, Rubus villosus* Rose Family	NOW COMBINED WITH *Rubus alleghaniensis*	189
MOUNTAIN BLACKBERRY *Rubus Allegheniensis* Rose Family	MOUNTAIN BLACKBERRY *Rubus alleghaniensis* Rose Family	192
LEAFY CLUSTER BLACKBERRY *Rubus argutus* Rose Family	HIGHBUSH BLACKBERRY *Rubus ostryifolius* Rose Family	192
THORNLESS BLACKBERRY *Rubus Canadensis* Rose Family	THORNLESS BLACKBERRY *Rubus canadensis* Rose Family	193
BLACK CHOKEBERRY *Aronia nigra, Pyrus arbutifolia,* Var. *melanocarpa* Apple Family	BLACK CHOKEBERRY *Aronia melanocarpa* Rose Family	193
OBLONG-FRUITED JUNEBERRY *Amelanchier oligocarpa* Apple Family	OBLONG-FRUITED JUNEBERRY *Amelanchier bartramiana* Rose Family	195
PORTER'S PLUM *Prunus Allegheniensis* Plum Family	ALLEGHANY PLUM *Prunus alleghaniensis* Rose Family	196
SLOE, BLACKTHORN *Prunus spinosa* Plum Family	SLOE, BLACKTHORN *Prunus spinosa* Rose Family	196
SAND CHERRY, DWARF CHERRY *Prunus pumila* Plum Family	SAND CHERRY, DWARF CHERRY *Prunus pumila* Rose Family	197

Peterson Nomenclature	Current Nomenclature	
WILD BLACK CHERRY, RUM CHERRY *Prunus serotina* Plum Family	BLACK CHERRY *Prunus serotina* Rose Family	198
BLACK CROWBERRY *Empetrum nigrum* Crowberry Family	NO CHANGE	202
INKBERRY, EVERGREEN WINTER BERRY *Ilex glabra* Holly Family	INKBERRY *Ilex glabra* Holly Family	203
BUCKTHORN *Rhamnus cathartica* Buckthorn Family	EUROPEAN BUCKTHORN *Rhamnus cathartica* Buckthorn Family	205
LANCE-LEAVED BUCKTHORN *Rhamnus lanceolata* Buckthorn Family	NO CHANGE	207
ALDER-LEAVED BUCKTHORN *Rhamnus alnifolia* Buckthorn Family	NO CHANGE	207
NORTHERN FOX GRAPE *Vitis Labrusca* Grape Family	NORTHERN FOX GRAPE *Vitis labrusca* Grape Family	208
SUMMER GRAPE *Vitis aestivalis* Grape Family	NO CHANGE	210
BLUE GRAPE *Vitis bicolor* Grape Family	SILVERLEAF GRAPE *Vitis argentifolia* Grape Family	211
RIVERSIDE OR SWEET-SCENTED GRAPE *Vitis vulpina, Vitis riparia* Grape Family	RIVERBANK GRAPE *Vitis riparia* Grape Family	212

Peterson Nomenclature	Current Nomenclature	
FROST OR CHICKEN GRAPE *Vitis cordifolia* Grape Family	FROST OR CHICKEN GRAPE *Vitis vulpina* Grape Family	214
VIRGINIA CREEPER, WOODBINE, AMERICAN IVY *Parthenocissus quinquefolia,* *Ampelopsis quinquefolia* Grape Family	VIRGINIA CREEPER *Parthenocissus quinquefolia* Grape Family	217
ANGELICA TREE, HERCULES' CLUB *Aralia spinosa* Ginseng Family	DEVIL'S-WALKINGSTICK *Aralia spinosa* Ginseng Family	218
AMERICAN SPIKENARD, INDIAN ROOT *Aralia racemosa* Ginseng Family	NO CHANGE	219
WILD OR VIRGINIAN SARSAPARILLA *Aralia nudicaulis* Ginseng Family	NO CHANGE	221
BRISTLY SARSAPARILLA, WILD ELDER *Aralia hispida* Ginseng Family	NO CHANGE	224
TUPELO, SOUR GUM, PEPPERIDGE *Nyssa sylvatica* Dogwood Family	BLACK TUPELO *Nyssa sylvatica* Tupelo (Sour Gum) Family	226
ALPINE OR BLACK BEARBERRY *Mairania alpina,* *Arctostaphylos alpina* Heath Family	ALPINE OR BLACK BEARBERRY *Arctostaphylos alpina* Heath Family	228

Peterson Nomenclature	Current Nomenclature	
BLACK OR HIGH-BUSH HUCKLEBERRY	BLACK OR HIGH-BUSH HUCKLEBERRY	229
Gaylussacia resinosa	*Gaylussacia baccata*	
Huckleberry Family	Heath Family	
DWARF OR BUSH HUCKLEBERRY	DWARF OR BUSH HUCKLEBERRY	230
Gaylussacia dumosa	*Gaylussacia dumosa*	
Huckleberry Family	Heath Family	
LOW BLACK BLUEBERRY	LOW BLACK BLUEBERRY	231
Vaccinium nigrum	*Vaccinium nigrum*	
Huckleberry Family	Heath Family	
FRINGE TREE	FRINGETREE	232
Chionanthus Virginica	*Chionanthus virginicus*	
Olive Family	Olive Family	
PRIVET	NO CHANGE	233
Ligustrum vulgare		
Olive Family		
BLACK OR GARDEN NIGHTSHADE	BLACK OR GARDEN NIGHTSHADE	234
Solanum nigrum	*Solanum nigrum*	
Potato Family	Nightshade Family	
AMERICAN ELDER, SWEET ELDER	COMMON ELDER	236
Sambucus Canadensis	*Sambucus canadensis*	
Honeysuckle Family	Honeysuckle Family	
MAPLE-LEAVED VIBURNUM OR ARROWWOOD, DOCKMACKIE	MAPLELEAF VIBURNUM	237
Viburnum acerifolium	*Viburnum acerifolium*	
Honeysuckle Family	Honeysuckle Family	
DOWNY-LEAVED ARROWWOOD	DOWNY-LEAVED VIBURNUM	239
Viburnum pubescens	*Viburnum rafinesquianum*	
Honeysuckle Family	Honeysuckle Family	

Peterson Nomenclature	Current Nomenclature	
WITHE-ROD *Viburnum cassinoides* Honeysuckle Family	WITHEROD, WILD RAISIN *Viburnum cassinoides* Honeysuckle Family	240
LARGER WITHE-ROD *Viburnum nudum* Honeysuckle Family	POSSUMHAW *Viburnum nudum* Honeysuckle Family	243
SWEET VIBURNUM, SHEEPBERRY, NANNY BERRY *Viburnum Lentago* Honeysuckle Family	NANNYBERRY *Viburnum lentago* Honeysuckle Family	244
BLACK HAW, STAG BUSH *Viburnum prunifolium* Honeysuckle Family	BLACKHAW *Viburnum prunifolium* Honeysuckle Family	246
COMMON JUNIPER *Juniperus communis* Pine Family	COMMON JUNIPER *Juniperus communis* Cedar Family	249
RED CEDAR *Juniperus Virginiana* Pine Family	EASTERN REDCEDAR *Juniperus virginiana* Cedar Family	252
SHRUBBY RED CEDAR *Juniperus Sabina* Pine Family	SHRUBBY REDCEDAR *Juniperus sabina* Cedar Family	254
YELLOW CLINTONIA *Clintonia borealis* Lily-of-the-Valley Family	CORN-LILY, BLUEBEARD-LILY *Clintonia borealis* Lily Family	257
BLUE COHOSH, PAPOOSE ROOT *Caulophyllum thalictroides* Barberry Family	NO CHANGE	257
SASSAFRAS *Sassafras sassafras, Sassafras officinale* Laurel Family	SASSAFRAS *Sassafras albidum* Laurel Family	258

Peterson Nomenclature	Current Nomenclature	
ROUND-LEAVED CORNEL OR DOGWOOD	ROUND-LEAVED DOGWOOD	262
Cornus circinata	*Cornus circinata*	
Dogwood Family	Dogwood Family	
SILKY CORNEL, KINNIKINNIK	SILKY DOGWOOD	263
Cornus Amonum, Cornus sericea	*Cornus amomum*	
Dogwood Family	Dogwood Family	
ALTERNATE-LEAVED CORNEL OR DOGWOOD	PAGODA DOGWOOD	266
Cornus alternifolia	*Cornus alternifolia*	
Dogwood Family	Dogwood Family	
BLUE TANGLE, TANGLEBERRY, DANGLEBERRY	DANGLEBERRY	269
Gaylussacia frondosa	*Gaylussacia frondosa*	
Huckleberry Family	Heath Family	
BOX HUCKLEBERRY	BOX HUCKLEBERRY	270
Gaylussacia brachycera	*Gaylussacia brachycera*	
Huckleberry Family	Heath Family	
GREAT BILBERRY	ALPINE BILBERRY	271
Vaccinium uliginosum	*Vaccinium uliginosum*	
Huckleberry Family	Heath Family	
DWARF BILBERRY	DWARF BILBERRY	272
Vaccinium caespitosum	*Vaccinium cespitosum*	
Huckleberry Family	Heath Family	
HIGH-BUSH OR TALL BLUEBERRY	HIGH-BUSH OR TALL BLUEBERRY	272
Vaccinium corymbosum	*Vaccinium corymbosum*	
Huckleberry Family	Heath Family	
DWARF BLUEBERRY	DWARF BLUEBERRY	277
Vaccinium Pennsylvanicum	*Vaccinium laevifolium*	
Huckleberry Family	Heath Family	

Peterson Nomenclature	Current Nomenclature	
LOW BLUEBERRY *Vaccinium vacillans* Huckleberry Family	LOW BLUEBERRY *Vaccinium vacillans* Heath Family	278
ARROWWOOD *Viburnum dentatum* Honeysuckle Family	SOUTHERN ARROWWOOD *Viburnum dentatum* Honeysuckle Family	279
SOFT-LEAVED ARROWWOOD *Viburnum molle* Honeysuckle Family	KENTUCKY VIBURNUM *Viburnum molle* Honeysuckle Family	282
BLUE OR MOUNTAIN FLY HONEYSUCKLE *Lonicera coerulea* Honeysuckle Family	BLUE OR MOUNTAIN FLY HONEYSUCKLE *Lonicera villosa* Honeysuckle Family	282
NORTH AMERICAN PAPAW *Asimina triloba* Custard-Apple Family	NO CHANGE	287
MAY APPLE, MANDRAKE, UMBRELLA LEAF, WILD LEMON *Podophyllum pelatum* Barberry Family	NO CHANGE	288
DWARF THORN *Crataegus uniflora, Crataegus parvifolia* Apple Family	ONE-FLOWER HAWTHORN *Crataegus parvifolia* Rose Family	290
DWARF GINSENG, GROUNDNUT *Panax trifolium, Aralia trifolia* Ginseng Family	DWARF GINSENG, GROUNDNUT *Panax trifolius* Ginseng Family	291
DEERBERRY, SQUAW HUCKLEBERRY *Vaccinium stamineum* Huckleberry Family	DEERBERRY, SQUAW HUCKLEBERRY *Vaccinium stamineum* Heath Family	292

Peterson Nomenclature	Current Nomenclature	
PERSIMMON, DATE PLUM *Diospyros Virginiana* Ebony Family	PERSIMMON *Diospyros virginiana* Ebony Family	293
LOW HAIRY GROUND CHERRY *Physalis pubescens* Potato Family	LOW HAIRY GROUND-CHERRY *Physalis pubescens* Nightshade Family	294
CUT-LEAVED GROUND CHERRY *Physalis angulata* Potato Family	CUT-LEAVED GROUND-CHERRY *Physalis angulata* Nightshade Family	296
CLAMMY GROUND CHERRY *Physalis heterophylla,* *Physalis Virginiana* Potato Family	CLAMMY GROUND-CHERRY *Physalis heterophylla* Nightshade Family	297
HORSE NETTLE *Solanum Carolinense* Potato Family	HORSE NETTLE *Solanum carolinense* Nightshade Family	298
GREEN ARROW-ARUM *Peltandra Virginica,* *Peltandra undulata* Arum Family	GREEN ARROW-ARUM *Peltandra virginica* Arum Family	301
CHOKEPEAR *Pyrus communis* Apple Family	PEAR *Pyrus communis* Rose Family	302
AMERICAN CRAB APPLE *Malus coronaria, Pyrus coronaria* Apple Family	SWEET CRAB APPLE *Malus cornaria* Rose Family	303
BAYBERRY, WAXBERRY *Myrica Carolinensis,* *Myrica cerifera* Bayberry Family	BAYBERRY, WAXBERRY *Myrica pensylvanica* Bayberry Family	307
WHITE MULBERRY *Morus alba* Mulberry Family	NO CHANGE	308

Peterson Nomenclature	Current Nomenclature	
SMALL MISTLETOE *Razoumofskya pusilla,* *Arceuthobium pusillum* Mistletoe Family	DWARF MISTLETOE *Arceuthobium pusillum* Mistletoe Family	309
AMERICAN MISTLETOE *Phoradendron flavescens* Mistletoe Family	NO CHANGE	310
WHITE BANEBERRY *Actaea alba* Crowfoot Family	NO CHANGE	313
POISON SUMAC *Rhus Vernix, Rhus venenata* Sumac Family	POISON-SUMAC *Toxicodendron vernix* Sumac Family	314

POISON, CLIMBING, OR
  THREE-LEAVED IVY
Rhus radicans, Rhus
  Toxicodendron
Sumac Family
	POISON-IVY
	
	Toxicodendron radicans
	
	Sumac Family
		315

RED-OSIER CORNEL OR DOGWOOD *Cornus stolonifera* Dogwood Family	RED-STEMMED DOGWOOD *Cornus stolonifera* Dogwood Family	317
PANICLED CORNEL *Cornus candidissima, Cornus* *paniculata* Dogwood Family	PANICLED DOGWOOD *Cornus racemosa* Dogwood Family	318
CREEPING SNOWBERRY *Chiogenes hispidula, Chiogenes* *serpyllifolia* Huckleberry Family	CREEPING SNOWBERRY *Gaultheria hispidula* Heath Family	320
SNOWBERRY *Symphoricarpos racemosus* Honeysuckle Family	SNOWBERRY *Symphoricarpos albus* Honeysuckle Family	323

RED OR REDDISH PURPLE

RED OR REDDISH PURPLE

AMERICAN YEW

Taxus minor. Taxus Canadensis   **Yew Family**

Fruit. — The fruit is drupelike; the hard, bony, dark-colored, oval seed being nearly inclosed in a red, pulpy cup, which is the developed fleshy flower disk. The drupe is solitary, growing at the end or the side of the branches. It is bracted at the base.

Leaves. — The leaves are about half an inch long, pointed, and green on both sides. They are arranged spirally around the branches.

Flowers. — The flowers are mostly dioecious. The fertile ones are solitary, and the sterile ones consist of a few naked stamens. April, May.

This low shrub has spreading, crooked branches. It delights in a shaded situation, especially favoring the shelter of evergreens. It is sometimes called Ground Hemlock from its resemblance to young hemlock growths. The

3

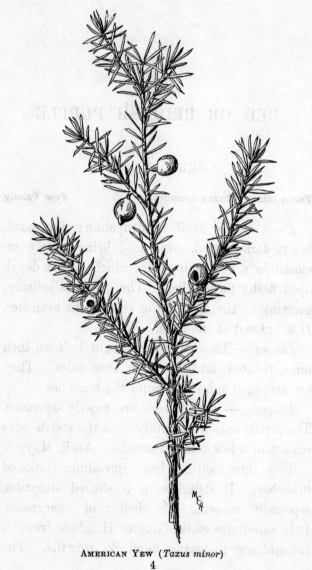

AMERICAN YEW (*Taxus minor*)

4

wood of the Yew is tough and elastic, and was used by the Indians in making their bows.

It extends south to New Jersey and along the Alleghanies to Virginia. It also ranges northward from Minnesota and Iowa.

INDIAN TURNIP.   JACK-IN-THE-PULPIT

Arisæma triphyllum   **Arum Family**

Fruit. — Bright, shining, scarlet berries are crowded together in an ovoid head. Each fruit bears the tip of the stigma at the top. One or two seeds are embedded in a scant, juicy pulp. August.

Leaves. — One or two three-parted leaves usually overtop flower and fruit. The leaflets are ovate and mostly entire. The leaves sometimes wither and fall before the fruit develops.

Flowers. — The flowers are borne at the base of a club-shaped spadix which is nearly inclosed in a sheathing spathe, the top portion of which curves over, forming a sheltering roof. The flowers are mostly dioecious, although one plant sometimes bears both staminate and pistillate flowers. They are fertilized by small insects

which crawl around within the sheathing spathe
and cover themselves with pollen.

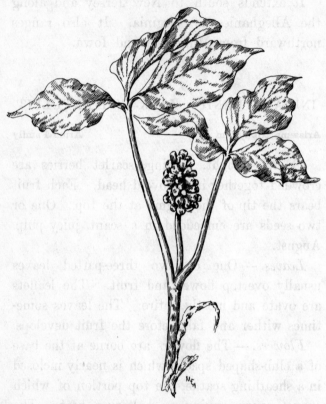

JACK-IN-THE-PULPIT (*Arisæma triphyllum*)

The plant has a turnip-shaped, wrinkled, pep-
pery corm, which contains much starch.   The

Indians are said to have cooked it for food. They also cooked and ate the berries.

The color of the hood and its markings is very variable, sometimes light green with light markings, and sometimes dark with purple stripes. I was interested, one spring, to see if the light green spathes really did inclose male flowers and the dark ones female, as some authorities think probable. I gathered, one day, about thirty-six specimens. Of twenty dark ones, fourteen were pistillate and six staminate. Of sixteen light ones, five were pistillate and eleven staminate. The majority seem to bear out the supposition. Another interesting matter came to my attention from the examination of the specimens. Sixteen plants each had two leaves, and of these fifteen were mostly pistillate and the other one had about as many staminate as pistillate blossoms. Of twenty specimens with one leaf, sixteen were mainly staminate and four pistillate. Later observations tend to show that the two-leaved specimens usually have pistillate flowers, and the one-leaved staminate.

Jack-in-the-pulpit loves rich, wet woods, and extends as far west as Minnesota and eastern Kansas.

GREEN DRAGON.   DRAGON ROOT

Arisæma Dracontium   **Arum Family**

Fruit. — The orange-red berries grow in a large ovoid head. They are one- to few-seeded.

Leaves. — The usually solitary compound leaf has nine to eleven radiating leaflets with the two side ones somewhat lobed. The leaflets are pointed and oblong-lanceolate.

Flowers. — Both staminate and pistillate flowers usually grow on the same spadix. The upper part of the spadix is much prolonged and extends considerably beyond the pointed spathe. Both spathe and spadix are green.

The range is about that of Jack-in-the-pulpit. Around the main club-shaped bulb, cluster many tiny bulblets, producing an effect which to a strong imagination might suggest the foot and toes of a monster and be responsible for the common name of Dragon Root. The radiating leaflets are a bit suggestive of a dragon's claws. Leaves and flowers are both green, the only bit of gay coloring that the plant affords appearing in the bright fruit cluster.

WATER ARUM

Calla palustris   **Arum Family**

Fruit. — The few red berries grow in an oblong head. The seeds are few and inclosed in jelly. The spathe is persistent in fruit. July, August.

Leaves. — The leaves are heart-shaped, and borne on erect or spreading stems.

Flowers. — The flower stem is nearly as long as those of the leaves. The open, spreading spathe has a white upper surface and is green beneath. The spadix is much shorter than the spathe, and is covered with flowers, the lower of which are perfect and the upper ones often staminate.

It is a low plant, less than a foot in height, and resembles the cultivated Calla. Spreading by a slender, creeping rootstock, it often occurs in masses. The rhizomes are used by the Laplanders in making a kind of bread. It flourishes in cold bogs in Virginia, Wisconsin, and Iowa, and northward.

ASPARAGUS

Asparagus officinalis　　　　　**Lily-of-the-Valley Family**

Fruit. — The red berries are globose, and about
as large as a small huckleberry. They are solitary or in pairs, and grow on a slender, jointed
stem from the axil of a scale, which is really a
modified leaf. The berry is three-celled, with
two seeds in each cell. The calyx lobes are at
the base of the berry. August, September.

Leaves. — The true leaves appear as scales
along the stem and branches. From the axils,
along the branches, grow three tiny threadlike
branchlets which are often mistaken for leaves.

Flowers. — The flowers are small, bell-shaped,
and greenish yellow. They grow on drooping,
jointed pedicels. June.

The Asparagus was introduced from Europe,
and has become quite a frequent roadside escape.
It is very attractive in fruit, making one think
of a miniature Christmas tree, with its gay
decorations of red balls. The thick shoots of
spring are edible, and bear the true leaves as
large scales, which persist on the base of the
plant until fall.

WILD SPIKENARD

Vagnera racemosa. Smilacina racemosa
Lily-of-the-Valley Family

Fruit. — The berries grow in a long racemose cluster at the terminus of the leafy, unbranched stem. They are globular, and when fully ripe, in late September, are translucent and a dull red in color. Before this, they present a peculiarly speckled appearance, being whitish, with many red dots and splashes. The flesh is thin. While the ovary is three-celled, with two ovules in each, the developed berry contains but one or two large seeds. The fruits have an aromatic flavor. September.

Leaves. — The leaves are alternate, nearly stemless, and have tiny hairs along the entire wavy margins. Each is oval-lanceolate, with a long, tapering point. They are so arranged along the stem that the plane of the upper surface is nearly parallel with the drooping stem, thus exposing it most advantageously to the light.

Flowers. — The small, white, six-parted flowers grow in terminal, pyramidal clusters. May, July.

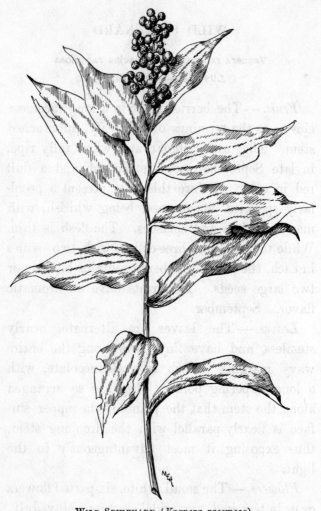

WILD SPIKENARD (*Vagnera racemosa*)

12

Both in flower and fruit this plant lends its attractiveness to the woodside road. The stem is zigzag and somewhat inclined. The rootstock is stout. It extends south to Georgia and west to Missouri and Arizona.

THREE–LEAVED SOLOMON'S SEAL

Vagnera trifolia. Smilacina trifolia

Lily-of-the-Valley Family

Fruit. — The globular berries grow in a few-fruited raceme. They are red when ripe.

Leaves. — The leaves are usually three in number, although two or four occasionally occur. They are oblong, and taper to a narrowed sheathing base. They are acute at the apex.

Flowers. — The white, six-parted flowers are smaller than in *V. stellata.*

This is a smaller plant than False Spikenard or Star-flowered Solomon's Seal. A casual observer might confuse it with *Unifolium Canadense,* but it may be distinguished from it by the narrowed sheathing base of its leaf and its six-parted flowers. It grows in bogs and wet woods in New England, New Jersey, Pennsylvania, and Michigan.

FALSE LILY–OF–THE–VALLEY. TWO-LEAVED SOLOMON'S SEAL

Unifolium Canadense.   Maianthemum Canadense
Lily-of-the-Valley Family

Fruit. — The berry is whitish, thickly speckled with red until late in the season, when it becomes a dull red. The fruits grow in a terminal cluster, and are much like those of *Vagnera trifolia.*

Leaves. — The leaves are ovate to lanceolate, with a heart-shaped base. There are usually two, sometimes three on the stem. They are sessile or nearly so.

Flowers. — The four-parted, small, white flowers grow in a simple raceme.

The heart-shaped base of the leaf and the four-parted perianth are "earmarks" of the species. This is a tiny plant, growing profusely in woods, sometimes in patches, sometimes alone. It is quite common throughout southern Canada and south to North Carolina, Iowa, and South Dakota.

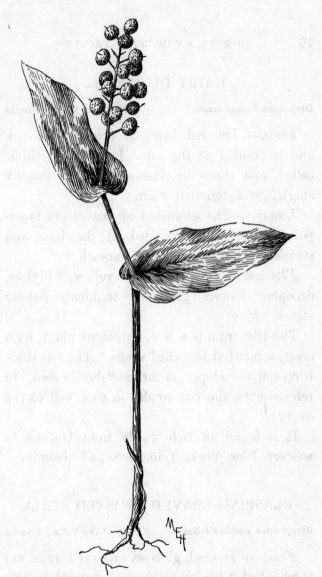

FALSE LILY-OF-THE-VALLEY (*Unifolium Canadense*)

15

HAIRY DISPORUM

Disporum lanuginosum   **Lily-of-the-Valley Family**

Fruit. — The red berry is oblong or ovoid, and is pointed at the top. It is pulpy, three-celled, and three- to six-seeded. It is usually single, on a terminal stem.

Leaves. — The ovate-oblong leaves are taper-pointed, somewhat rounded at the base, and stemless. They are downy beneath.

Flowers. — The greenish yellow, lilylike, drooping flowers grow on terminal, slender stems. May.

The Disporum is a low, pubescent plant, with erect, somewhat branched stems. The rootstock is creeping. *Disporum* means "double seed," in reference to the two ovules in each cell of the ovary.

It is found in rich woods from Ontario to western New York, Tennessee, and Georgia.

CLASPING–LEAVED TWISTED STALK

Streptopus amplexifolius   **Lily-of-the-Valley Family**

Fruit. — The red, globose, or oval berries are three-celled, with many seeds, arranged in two

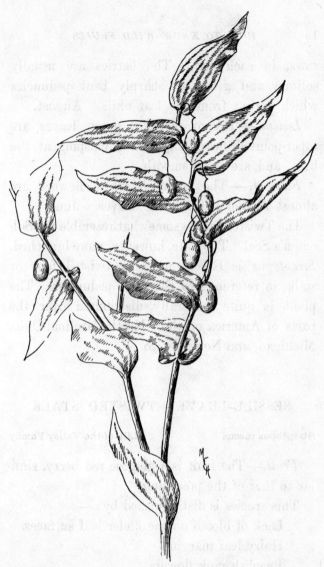

CLASPING-LEAVED TWISTED STALK (*Streptopus amplexifolius*)

rows, in each cell. The berries are usually solitary and grow on sharply bent peduncles which spring from the leaf axils. August.

Leaves. — The ovate, light green leaves are taper-pointed, heart-shaped, and clasping at the base, and are very smooth.

Flowers. — The greenish white flowers are almost hidden beneath the leaves. June.

The Twisted Stalks somewhat resemble the Solomon's Seal. They are, however, more branched. *Streptopus* is Greek, meaning twisted foot or stalk, in reference to the bent peduncles. The plant is quite generally distributed over the parts of America north of North Carolina, Ohio, Michigan, and New Mexico.

SESSILE–LEAVED TWISTED STALK

Streptopus roseus   **Lily-of-the-Valley Family**

Fruit. — The fruit is a globose red berry, similar to that of the preceding.

This species is distinguished by : —

Lack of bloom on the under leaf surfaces.
Hairy leaf margins.
Purplish pink flowers.

Less abruptly bent flower stalks.

Earlier period of blooming.

Georgia, Michigan, and Oregon are its southern limits.

SESSILE–FLOWERED WAKE–ROBIN

Trillium sessile   **Lily-of-the-Valley Family**

Fruit. — The red, stemless berry is globular and about half an inch long.   It is six-angled, three-celled, and many-seeded.

Leaves. — The three whorled leaves are likewise sessile.   They are ovate, with acute tips, and are often spotted with lighter and darker green.

Flowers. — The sessile flowers are a dull red, occasionally greenish.   They have narrow sepals and petals, and an agreeable odor.   April, May.

This plant of moist woods grows in Pennsylvania and southward to Florida.   Minnesota and Arkansas are its western limits.

EARLY WAKE–ROBIN

Trillium nivale   **Lily-of-the-Valley Family**

Fruit. — The round, flattened berry is small, about a third of an inch in diameter.   It is red and has but three rounded divisions. This and *Trillium erythrocarpum* are the only three-angled *Trillium* fruits. The berry is short-stemmed.

Leaves. — The three ovate, whorled leaves are blunt at the apex, and have short stems.

Flowers. — The flowers are small and white, with erect, spreading petals. March, May.

This is a dwarf species, only two to five inches high. It grows in woods from Pennsylvania and Ohio, south to Tennessee and Iowa.

LARGE–FLOWERED WAKE–ROBIN

Trillium grandiflorum　　　　　　**Lily-of-the-Valley Family**

Fruit. — The red berry is slightly six-angled and from one-half inch to an inch in length. The sepals persist at the base as do also the filaments, which remain green. The berry is globular, three-celled, and many-seeded. The peduncle is sometimes three inches long.

Leaves. — Somewhat four-sided, but not as broad as *Trillium erectum*. They are pointed and nearly stemless — in the usual whorl of three.

Flowers. — The flowers are large, with erect

and spreading white petals that later grow pink. They grow on long, erect stems.

This is a native of rich woods from Vermont to North Carolina, with Minnesota and Missouri as the western boundary of its range.

ILL–SCENTED WAKE–ROBIN. BIRTHROOT

Trillium erectum   **Lily-of-the-Valley Family**

Fruit. — The dark red, round-ovate berry is distinctly six-angled. A dry stigma is at the junction of each two of these angles or ridges. The persistent dry sepals and remnants of the petals are at the base. The fruit is borne on a long, somewhat reclined stem. The berry is about an inch long. The brown seeds are numerous and horizontal in each cell. August.

Leaves. — The leaves are broadly four-sided, with scarcely any stems. The apex is acute. The leaves grow in a whorl of three at the summit of the plant stem. At the time of fruiting they are apt to be torn, faded, and brown.

Flowers. — The terminal, solitary flower is somewhat reclined and varies much in color;

ILL-SCENTED WAKE-ROBIN (*Trillium erectum*)

white, pink, dark red, yellow, and even greenish
blossoms having been found. The odor is very
unpleasant. April, June.

When driving in early spring along a wooded roadway, I have found great patches of this curiously colored blossom growing near streams or in swampy grounds. I was quite content, however, to admire them from a distance, objecting to the odor which a closer acquaintance entails.

This odor and the color of the flower serve the plant a useful purpose in attracting the flesh fly, which Clarence M. Weed says is the most useful insect in disseminating the pollen of this plant. The color of the flower resembles that of raw meat, and the yellow specimens which I saw were quite the color of fat beef.

This species grows as far west as Minnesota and Missouri, and south to North Carolina.

NODDING WAKE-ROBIN

Trillium cernuum   **Lily-of-the-Valley Family**

Fruit. — The reddish ovate berry is somewhat six-angled, and is borne on a short inclined or recurved stem. It is about three-quarters of an inch long.

Leaves. — The three leaves are sessile or nearly so. They are very similar to *Trillium erectum*.

Flowers. — The white or pink flowers have wavy-margined petals that roll backwards. They grow on a curved stem and are often hidden beneath the leaves. April–June.

This Wake-robin favors rich woods, and ranges from Ontario to Georgia and Missouri and west to Minnesota.

PAINTED WAKE–ROBIN

Trillium undulatum.   Trillium erythrocarpum
Lily-of-the-Valley Family

Fruit. — The bright red berry is about three quarters of an inch long, and is borne terminally on a nearly erect stem. It is ovate, with the narrow end rather pointed, is obscurely three-angled, and crowned with the persistent stigmas. Its three angles instead of six, lack of wings, and ovate shape distinguish it from all the other *Trillium* fruits. The three spreading sepals are persistent at the base as are also dry remnants of the petals. The skin is thin, the pulp white and scanty, and the brown seeds numerous, ovate, and arranged horizontally.

Leaves. — The leaves are in a whorl of three. Each leaf is petioled. The stems unite, forming a triangular surface, from the center of which springs the flower stem. The leaves are broad-ovate, with a long tapering point.

Flowers. — The recurved white petals are marked with crimson stripes at the base. This Wake-robin bears one of the most beautiful flowers of the genus.

The Painted Trillium grows in profusion in the Catskill and Adirondack mountains, and is found in damp woods as far west as Wisconsin and Missouri and south to Georgia.

WALTER'S GREENBRIER

Smilax Walteri   **Smilax Family**

Fruit. — The coral-red, globose berries are in umbels, growing on flattened stems which scarcely equal the petioles in length. The berries are two- to three-seeded.

Leaves. — The ovate or ovate-lanceolate leaves are thick and green on both sides. They are somewhat heart-shaped at base, but are seldom lobed. The apex is bristle pointed.

Flowers. — The blossoms are brownish, and grow in umbels. April–June.

This is the only Greenbrier in our section which bears red berries. *Smilax Walteri* has a low stem, somewhat prickly below and unarmed above. It grows in swamps and moist places as far north as New Jersey.

LAUREL MAGNOLIA

Magnolia Virginiana.   Magnolia glauca   Magnolia Family

Fruit. — The small conelike fruit consists of many coherent carpels, which are crowded upon the enlarged receptacle. When mature, the conelike mass is red and each carpel splits along its outer side. The one or two contained red seeds escape, but, for a time, each remains hanging by a slender white thread. The seeds are slightly bitter, but are used as food by the birds. September, October.

Leaves. — The oval or elliptical leaves have a leathery appearance. They are light green and shining above and much whitened beneath. In the South, they usually remain on the tree during the winter, falling in the spring to give place to new growths. The petioles are short and tapering.

LAUREL MAGNOLIA (*Magnolia Virginiana*)

27

Flowers. — Emerson says : "The flower, two or three inches broad, is as beautiful and almost as fragrant as a water lily." It is creamy white, solitary, and terminal. June.

The soft white flowers, amidst the glossy green foliage, yield one of the pleasures found among the swamp growths of early summer, especially along the coast. The gradual transition from one part of the flower to another is interesting. The sepals are much like the petals and the stamens retain a petal-like character. The fruit mass flies the red-seed banners to attract the bird carriers. In our section the plant is a shrub; along the Gulf of Mexico it becomes a tree. It has been found as far north as Cape Ann. The bark is usually brown, but on young growths is light gray.

CUCUMBER TREE. MOUNTAIN MAGNOLIA

Magnolia acuminata   **Magnolia Family**

Fruit. — The structure of the fruit is similar to that of the small Magnolia, but it is much larger. It is pink with red seeds. The resemblance of the green fruit to a cucumber is the cause of one of its common names.

The fragrant flowers are greenish white and rather inconspicuous, being so nearly the color of the foliage.

This plant is rare in our section. It occurs in western New York and southward. It is a tree sixty to ninety feet high, with brown bark.

YELLOW PUCCOON. YELLOW ROOT YELLOW INDIAN PAINT. ORANGE ROOT. GOLDEN SEAL

Hydrastis Canadensis   **Crowfoot Family**

Fruit. — The fruit somewhat resembles a raspberry. It is a small head of one- to two-seeded crimson berries. The head is ovoid and blunt, and the fleshy carpels are tipped with short, curved beaks.

Leaves. — There is a single roundish root leaf and near the top of the stem are two more rounded leaves. These are five- to seven-lobed, doubly serrate, and heart-shaped at the base.

Flowers. — The blossoms are borne at the top of the stem. They are greenish white and inconspicuous. The sepals fall when the flower

RED BANEBERRY (*Actæa rubra*)

32

opens. There are no petals, just numerous stamens and several pistils.

This is a low, hairy perennial about a foot high. Its rootstock is thick, knotted, and yellow. The plant grows in rich woods from New York to Minnesota and southward.

RED BANEBERRY

Actæa rubra. Actæa spicata, Var. rubra

Crowfoot Family

Fruit. — The cherry-colored, oval berries grow in terminal ovate clusters about three inches in length. A white berry is occasionally found. Each fruit is borne on a *slender* stem, and has a groove along one side extending from the stem to a black spot at the opposite end, the remnant of the stigma. The flesh of the berry is white and rather thin. The seeds are smooth and packed in two horizontal rows, with the points of attachment to the flesh on the grooved side. This fruit is a good illustration of the development of a simple pistil. The seeds differ in shape according to the position each occupies in the row. July, August.

Leaves. — This perennial bears two or three twice- or thrice-compound leaves. The leaflets are often lobed and sometimes the lower end ones are compound. They are coarsely toothed.

Flowers. — The small white flowers grow in terminal ovate clusters. The sepals fall when the flower opens. The stamens, protruding beyond the petals, give the raceme a feathery appearance. The stigma, maturing before the anthers shed their pollen, necessitates cross fertilization, which is effected by small bees.

This is a plant of the woods, and the fruit, with its beautiful rich coloring, brightens the wooded roadside in July. Near by the Wild Sarsaparilla drupes are blackening, while in more open spaces the Red and the Black Raspberries offer their delicious fruits.

Its range is from Maine, south to New Jersey and Pennsylvania, and westward.

COMMON BARBERRY

Berberis vulgaris   **Barberry Family**

Fruit. — The oblong scarlet berries grow in clusters, which are usually drooping. Each berry commonly has one seed, which is erect and

COMMON BARBERRY (*Berberis vulgaris*)

35

is covered with a hard, brittle coat.   The fruit is very acid but is eatable when cooked.   It makes a delicious jelly.   The berries are eaten by birds and the seeds thrown up from the crop instead of passing through the entire digestive tract. September.

Leaves. — The leaves seem to grow in rosettes from the axils of the spines.   They are oval to obovate and bristly toothed.

This spiny shrub generally grows in thickets and waste grounds in eastern New England, having become thoroughly wild there.   It varies in height from one to six feet.   The wood and inner bark are yellow.   The spines, in groups of seven or three, are modified leaf structures and protect, against destruction from grazing animals, the fresh shoots, with their leaves or flowers, which grow from their axils.

The flowers show ingenious arrangements for protecting the pollen against dew or rain, and for securing cross fertilization.   The yellow blossoms grow in drooping, many-flowered racemes, and the concave petals of each bloom thus act as a roof for the pollen borne by the stamens which they cover.

The lower third of each stamen is sensitive to

the slightest touch. Both the hive bee and the humblebee come to the flower in quest of the honey which is produced by the saffron-colored swellings on the petals and in getting it are almost sure to touch the base of the stamens. These spring up and cover the head and parts of the fore legs and proboscis of the bee with pollen.

> " All down the loose-walled lanes in archin' bowers,
> The barb'ry droops its strings o' golden flowers,
> Whose shrinkin' hearts the school-girls love to try
> With pins, — they'll worry yourn so, boys, bimeby ! "
> — LOWELL'S *Sunthin' in the Pastoral Line.*

The peasants of Europe, long before science explained the phenomenon, declared that barberry bushes caused wheat to rust. The fungus causing wheat rust often lives but part of its life on wheat. There is one stage of its growth which takes place on leaves of the barberry. Its presence there is manifested by groups of little orange-colored cups, called " cluster cups," which grow on the under surface of the leaf.

The state legislature of Massachusetts, as early as 1760, passed —

" An Act to prevent Damage to English Grain arising from Barberry Bushes."

There are certain grasses, however, upon which the " summer spore " stage of the " wheat rust " is produced throughout the year. This stage does not need, in the spring, the intermediate host of the barberry leaf, but will grow directly on the young grain. Eradication of the barberry, therefore, while necessary, and of advantage, does not always eradicate the trouble.

WILD ALLSPICE.   BENJAMIN BUSH
FEVER BUSH.   SPICE BUSH

Benzoin Benzoin.   Lindera Benzoin   **Laurel Family**

Fruit. — The oval drupes are red and shining, with thin, yellow flesh and a large stone. They grow in small bunches, of from two to five, on stout, short stalks. September.

Leaves. — The oval or elliptical leaves are short-pointed at the apex and narrowed at the base. The under surface is paler than the upper. Yellow is the fall color.

Flowers. — The flowers are small, yellow, and usually dioecious. They open before the leaves appear in the spring. April, May.

This is one of the early blooming spring shrubs. The flowers and leaves, especially if

SPICE BUSH (*Benzoin Benzoin*)

39

bruised, have an aromatic odor which, to some, is very disagreeable. It belongs to a family which includes such plants as the Camphor and Cinnamon. Its home is in damp woods. North Carolina, Tennessee, and Kansas limit its southern range.

HAWTHORN OR NORTHERN GOOSE-BERRY

Ribes oxyacanthoides   **Gooseberry Family**

Fruit. — The reddish purple berry is round or round-ovoid, smooth, and covered with a bloom. Like the other species, it keeps the dried calyx at the summit, and has similar seeds. July, August.

Leaves. — The leaves are deeply three- to five-lobed, with the lobes toothed and cut. The base of the leaf is heart-shaped or wedge-shaped.

Flowers. — The greenish or purplish flowers grow in few-flowered clusters, on short pedicels.

This is a low, usually smooth, shrub, with crooked or reclined branches. When prickles occur, they are scattered, and the spines, if any, grow singly or in threes. The plant grows in wet woods as far south as New Jersey and west

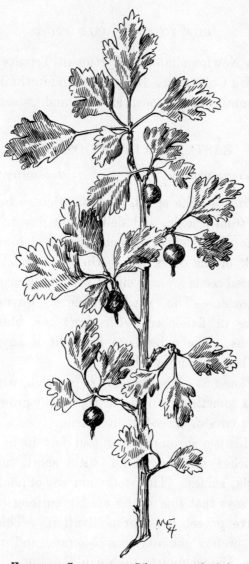

HAWTHORN GOOSEBERRY (*Ribes oxyacanthoides*)

41

from Newfoundland to Northwest Territory and
British Columbia. It is also found in the Rocky
Mountains as far south as Utah and Colorado.

EASTERN WILD GOOSEBERRY

Ribes rotundifolium   **Gooseberry Family**

Fruit. — The small, purplish, globose berry is
free from prickles and delicious in flavor. It is
borne on a slender, smooth stem and bears the
mark of the calyx at the tip. The gelatinous-
covered seeds are suspended in a pulpy mass.

Leaves. — The small, roundish leaves are
three- or five-lobed, with short and blunt divisions. The pubescence is slight if any, and
the leaf is shining above.

Flowers. — The flowers are greenish, with the
lobes sometimes a dull purple. They grow on a
short two- or three-flowered stem.

This is a shrub three or four feet high. The
branches are spreading, with short, usually
single, spines. The stems are smooth. Emerson says that this is the most promising of our
native gooseberries for cultivation. This species prefers mountainous *habitats*, and ranges
from Massachusetts to North Carolina.

SWAMP GOOSEBERRY

Ribes lacustre   **Gooseberry Family**

Fruit. — The berry of the Swamp Gooseberry is small, about one-sixth of an inch through. It is prickly, although the bristles are weak. It is reddish or dark purple, and often grows in raceme-like clusters. The dried calyx persists at the summit. The seeds have crustaceous coats, surrounded by gelatinous ones, and are suspended by tiny threads. The flavor is unpleasant. July, August.

Leaves. — One characteristic of the species is its deeply cut, five-lobed leaves. The petioles are slender and hairy. The leaf is thin, and hairy along the veins beneath.

Flowers. — The greenish flowers grow in many-flowered racemes, differing in this respect from the other gooseberries.

Ribes lacustre seems to be an intermediate form between gooseberries and currants. The young stems are quite prickly, and the spines are weak and single or clustered. The older branches are smooth, excepting a few axillary spines. The plant favors wet woods or swamps

from Pennsylvania north to Newfoundland, and
west through British Columbia and our northern
boundary states.

FETID CURRANT. MOUNTAIN CURRANT
PROSTRATE CURRANT

Ribes prostratum   **Gooseberry Family**

Fruit. — The round, light red berries grow in
slender racemes.   The berries and their short
stems are covered with glandular bristles.   When
bruised, they smell like Skunk Cabbage.

Leaves. — The leaves are deeply five- to seven-
lobed.   The lobes are ovate, acutish, and doubly
serrate.   The leaf stems are slender.

Flowers. — The greenish flowers grow in erect,
slender, several-flowered racemes.

Prostrate stems, sometimes rooting, are characteristic of this species.   The branches have
neither prickles nor spines.   The unpleasant
odor, when bruised, of both plant and fruit is
responsible for the name "Fetid Currant."   It
favors cold, wet woods, and extends south from
Labrador, especially along the Alleghanies, to
North Carolina, along the Rockies to Colorado,
and throughout southern Canada.

RED CURRANT

Ribes rubrum.   Ribes rubrum, Var. subglandulosum
Gooseberry Family

Fruit. — The smooth, round, red berries grow in drooping racemes, and in appearance and taste are similar to the cultivated currant.

The Wild Red Currant is very similar to the garden one, although that is probably a cultivated form of a European species. The currant is native to America, Europe, and Asia. The Red Currant and the Fetid Currant are the only species with red berries, and the Fetid is easily distinguished by its glandular bristles and disagreeable odor. The range of our native fruit is in cold woods, from Labrador to Alaska and south to New Jersey, Indiana, and Minnesota.

PURPLE–FLOWERING RASPBERRY

Rubus odoratus   **Rose Family**

Fruit. — The tiny red drupes are closely packed together into a flat, close head, which separates readily from the broad receptacle. Withered stamens and recurved calyx lobes

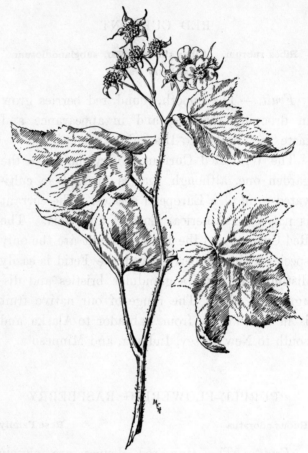

PURPLE-FLOWERING RASPBERRY (*Rubus odoratus*)

surround the base.   The fruits grow in clusters on bristly stems.   The "berries" are acid.

Leaves. — The large leaves are pubescent on both surfaces. They are from three- to five-lobed and finely toothed.

Flowers. — Large, purple, roselike blossoms grow in loose clusters.

It is for the attractiveness of the flowers rather than value of fruit that the Purple-Flowering Raspberry is known. The large peculiarly colored blossoms, so like the single rose in shape, appear advantageously against the large, soft, green leaves. " Grape leaves," Small Boy calls them, and they are quite similar. The shrub has no thorns, but recent growths; stems, leaves, and calyx are densely clothed with glandular hairs. The plant occurs in rocky woods as far south as Georgia and Tennessee and west to Michigan.

CLOUDBERRY. BAKED–APPLE BERRY
MOUNTAIN RASPBERRY

Rubus chamæmorus   **Rose Family**

Fruit. — The fruit consists of a few small drupes borne on a flat, broad receptacle, from which they separate when ripe. The flavor of the ripened fruit is pleasant, being sweet and honeylike. The ovate calyx lobes support the fruit at its base. It is yellow or amber-colored and usually tinged with red on the surface exposed to the sun. It is solitary and borne on a terminal stem.

Leaves. — Two, simple, roundish, five- to nine-lobed leaves, somewhat like geranium leaves, grow on the unbranched stems. They are serrate and alternate.

Flowers. — The blossoms are white. Staminate flowers grow on one plant; pistillate, on another.

This is a low herbaceous plant without prickles, which, in New England, is found along the coast of Maine and on the highest peaks of the White Mountains. It grows quite abundantly in Nova Scotia, Labrador, Newfoundland, and in the northern part of Quebec. It flourishes in greatest profusion even farther to the north, being an Arctic plant in Europe and Asia as well as in America. The northern berries are superior in size and quality.

The Indians in northern Quebec cook the berries in a sugar made from birch juice, and the dwellers in the posts of the Hudson Bay Company make from them a jam of rare flavor.

WILD RED RASPBERRY

(For illustration, see page 182.)

Rubus strigosus   **Rose Family**

Fruit. — The so-called berry is an aggregate fruit, consisting of many small, united drupes, the juicy pulp arising from the outer coat of the contained nutlet. The styles are persistent over the hemispherical surface of the fruit, and the persistent stamens surround the base. When ripe, the fruit separates from the white, spongy, oblong or conical receptacle. The fruits are borne in a loose cluster, either terminally or from a leaf axil. The fruit stems are thickly covered with recurved bristles. The fruits are red and delicious in taste and fragrance. July–September.

Leaves. — The compound leaves are composed of three or five leaflets. These are coarsely and irregularly serrate, and the lateral ones are sessile. They are rounded at base and acute at apex. The under surface is whitish and downy.

Flowers. — The white flowers grow in loose clusters.

Aside from its dissemination by seeds the Raspberry is spread from the root. Suckers run out in all directions from the central root and send up new shoots in fresh soil. It is common to find the Raspberry growing in patches by the roadside, along fence rows, or in corners.

Our Wild Red Raspberry is the ancestor of the various cultivated varieties. The cultivated White Raspberry is considered a "sport."

Rubus neglectus, or Purple Wild Raspberry, is an intermediate form between the Wild Red Raspberry and the Black Raspberry. It is a plant with comparatively few bristles or prickles. The fruit is borne on upright stems, is dark red, and nearly hemispherical. In cultivation the fruit is yellow.

DWARF RASPBERRY

Rubus Americanus.   Rubus triflorus   Rose Family

Fruit. — This fruit resembles in appearance that of the Low Blackberry. It differs in color, being dark red when ripe, and also in the separation of the few, two to five, grains which compose it from the receptacle, when the fruit is mature. Each grain is a juicy drupe inclosing a single hard-coated seed. The fruit is borne on a slender stem. July.

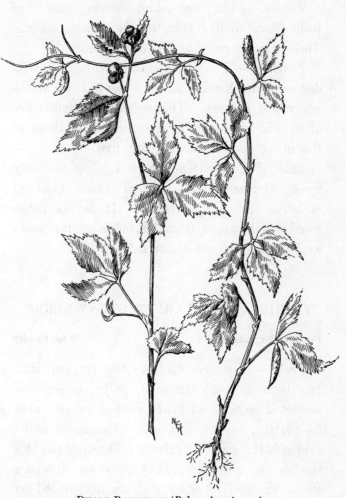

DWARF RASPBERRY (*Rubus Americanus*)

Leaves. — The compound leaves consist of from three to five thin, nearly smooth leaflets. These are coarsely and doubly serrate.

Flowers. — The flower cluster grows on a slender stem and consists of from one to three smallish white flowers. The sepals and petals are often six or seven in number, while those of the other species number but five.

This vine is ascending or trailing, slightly woody and hairy, but has no prickles. The fruit is borne on upright stems. It favors moist woods, and ranges from Labrador as far south as New Jersey and westward.

VIRGINIA OR SCARLET STRAWBERRY

Fragaria Virginiana   **Rose Family**

Fruit. — The receptacle of the ripened fruit has become much enlarged, pulpy, sweet, and scarlet in color; and bears, sunken in pits over its surface, several achenes. The lobed calyx subtends the aggregate fruit. The fruits of this species are globular. They grow on drooping stems, in small clusters, and are overtopped by the leaves.

Leaves. — The radical leaves consist of three broadly oval or obovate leaflets which are thick and leathery. The leaflets are obtuse, bluntly toothed, and hairy. The leaf stems are hairy, as are also the stipules at the base of the petioles.

Flowers. — The white flowers grow in small racemes on naked, hairy stems. They have many bright yellow stamens, which form a pleasing contrast with the white petals.

This is the common field strawberry of our section. The strawberries, aside from propagation by means of seeds, spread by runners, and the plants are usually found growing in patches. Fence corners, sandy knolls, and around rocks are spots which often reward our search for the berries. The common attractive color combination of red and green is seen in the leaves as well as in the leaves and fruit. At the time of fruiting some of the leaflets are often a bright red.

Nor do the fruits depend upon color alone as a means of allurement, but send forth upon the breezes a deliciously perfumed notice that they are ready for guests. Have you not encountered it and, following its lead, shared with the robins, bluebirds, and downy woodpeckers, the delicious feast ? The wild flavor of the berries is beyond the power of cultivation to produce or retain.

A strawberry bearing white fruits grows in the Alps.

F. Virginiana grows from New Brunswick southward and as far west as South Dakota.

Fragaria Canadensis or Northern Wild Strawberry, described as a separate species by Britton and Brown, is especially a northern plant. The leaflets are oblong or narrowly obovate, and have comparatively few teeth. The fruit is oblong or somewhat rounded at the summit. The achenes are sunken in pits.

EUROPEAN WOOD STRAWBERRY

Fragaria vesca   **Rose Family**

Fruit. — The achenes are not sunken in pits but are borne on the nearly smooth surface of the conical or hemispherical fruit. The calyx lobes are sometimes spreading, sometimes reflexed. The fruit cluster rises above the leaves.

Leaves. — The thin, light green, three-parted leaves grow on stems that are shorter than the flower stems.

This species belongs to fields and rocky places. It is naturalized from Europe, and is less common than *Fragaria Virginiana.*

Fragaria Americana, American Wood Strawberry, is by some considered a variety of *Fragaria vesca,* but is described by Britton and Brown as a distinct species. The leaflets are thinner and the fruit ovoid, or like a prolonged cone. The berry has a smooth, shining surface, looking almost as if varnished, and the achenes adhere but slightly to it. It is an inhabitant of rocky woods, and does not extend below Pennsylvania and New Jersey. Oregon is its western boundary.

THE ROSE

The rose, with its dainty pink coloring, and its subtle fragrance, is a general favorite. Both in blossom and in fruit it presents interesting features of structure. This is one of the plants that protects its pollen from rain and dew by pitching a petal tent over the stamens. You surely remember the overlapping, folded aspect of the petals in the early morning or on a cloudy day.

The rose produces no honey for the bee, but

does offer a liberal supply of pollen. The mass
of carpels at the center of the flower affords a
convenient landing place for the insect and a
substantial platform on which he may stand
while gathering the pollen stores, which are
yielded by the numerous stamens circled about.
During his harvesting the bee carries pollen
from one blossom to the receptive stigmas of
another, and accomplishes the cross fertilization
of the flower at the same time that he is gathering material for his " bee bread."

The fruit of the rose is peculiar to itself and
is known as a hip. It is considered by Gray
and by Britton and Brown to be a fleshy calyx
cup with a contracted mouth which incloses the
bony achenes. Kerner and Oliver consider the
hip as a hollow receptacle which contains carpels
that are entirely distinct from the wall of the
receptacle. The remnants of the styles remain
at the mouth of the hip, which may or may not
be surrounded by the calyx lobes.

The fruits are eaten by birds and the seeds
scattered by them. Mice, too, are fond of the
hips but gnaw and destroy the seeds instead of
aiding in their dispersal. Some rose hips were
gathered from the bushes and scattered along

the near-by path. In the morning, these were found to have been nibbled or eaten by the mice, while the hips on the bushes were left untouched, having been protected by the sharp thorns and prickles. These also hinder snails and caterpillars from reaching and destroying the fresh foliage.

There are five native species quite common in our section.

SMOOTH OR MEADOW ROSE

Rosa blanda   **Rose Family**

Fruit. — This globose, bright scarlet hip is generally smooth and retains the calyx lobes, which are erect on the fruit and somewhat hairy. September.

Leaves. — The leaflets are five to seven in number, obtuse at the summit, narrowed at the base, and simply and sharply serrate. They have short stems or are sessile. The stipules are broad and dilated.

Flowers. — The pink flowers are solitary or in corymbs.

Rosa blanda is a low bush not more than four feet high. It occasionally bears a few prickles

but entirely lacks spines, this feature being a
distinguishing mark of the species. The stems
are a dark red. It favors moist, rocky places
from Newfoundland south to New Jersey and
west to Illinois and Ontario.

SWAMP ROSE

Rosa Carolina   **Rose Family**

Fruit. — The scarlet hip is globose or de-
pressed-globose. It and the stem are set with
glandular hairs. The spreading or reflexed
calyx lobes are deciduous. September; remains
during the winter.

Leaves. — The compound leaves have five to
nine leaflets, usually seven. They are usually
narrowly oblong and pointed at either end.
They are simply and finely serrate, dull green,
and pale or pubescent beneath. Even in mid-
summer they often become a dull reddish color,
which is the regular autumnal shade. The
stipules are dilated.

Flowers. — The bright pink flowers usually
grow in corymbs, seldom solitary. June–
August.

This rose of swamps and stream borders

suckers freely and often grows in clumps. It is from one to eight feet in height. The spines are stout and often recurved. Prickles frequently occur along the stems. The range is throughout the Eastern United States. It is one of the most common species.

LOW OR PASTURE ROSE

Rosa humilis   **Rose Family**

Fruit. — The depressed globular or globose hips, with their pedicels, are hairy and glandular. The calyx lobes are not persistent. September and persistent.

Leaves. — The leaflets are from five to seven, usually five. They are coarsely serrate, rather thin, acute at apex, short-stemmed or sessile. The stipules are narrow and entire. Bright reds and orange are the autumnal colorings.

Flowers. — The solitary or two- to three-clustered pink flowers have a glandular calyx with lobed calyx lobes. May–July.

The Pasture Rose is usually low, about three feet high. The spines are slender and straight, and the bush is more or less prickly. It is a

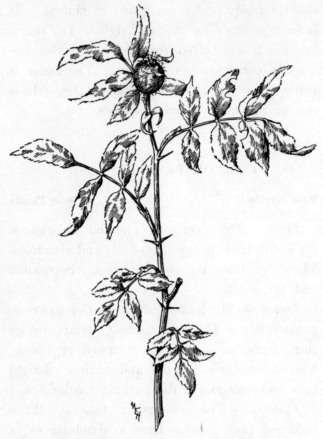

LOW OR PASTURE ROSE (*Rosa humilis*)

rose of dry soil and spreads rapidly by suckers. It extends south to Georgia and Louisiana, and west to Wisconsin.

NORTHEASTERN ROSE

Rosa nitida   **Rose Family**

Fruit. — The hip is globular, scarlet, and bears glandular hairs. The calyx lobes fall. Fruit is persistent.

Leaves. — The leaflets are usually narrowly oblong and pointed at either end. They are sharply serrate, bright green, and shining. The stipules are generally broad and somewhat glandular. Bright orange and red are the fall colors.

Flowers. — The flowers are in small clusters. June, July.

A marked sign of this species is its red shoots with their prickles, which are nearly as stout as the slender spines. It is a plant of low stature, about two feet in height. Its range is quite limited — from Massachusetts north to Newfoundland.

DOG ROSE. CANKER ROSE

Rosa canina   **Rose Family**

Fruit. — The reflexed calyx lobes fall from the long ovoid hip, which is usually smooth. The fruit is red when mature. September.

Leaves.—The stipules are glandular and broad. The five to seven leaflets are quite thick, nearly smooth above, somewhat pubescent below, and sharply toothed.

Flowers. — The flowers are often light pink or white. They are usually solitary, sometimes few-clustered.

This species is sometimes ten feet high. It has stout spines with hooks. It is similar to the following species, but is not fragrant. It has been naturalized from Europe. It frequents roadsides south to New Jersey and eastern Pennsylvania.

SWEETBRIER. EGLANTINE

Rosa rubiginosa   **Rose Family**

Fruit. — The ovoid hip changes from yellowish to red in ripening. It is usually smooth, sometimes slightly prickly with a prickly pedicel. The calyx lobes usually fall. September.

Leaves. — The leaflets are usually doubly and finely toothed. The under surface is densely hairy and resinous. The apex is generally obtuse and the base rounded. The leaf stems are prickly and the stipules are broad and glan-

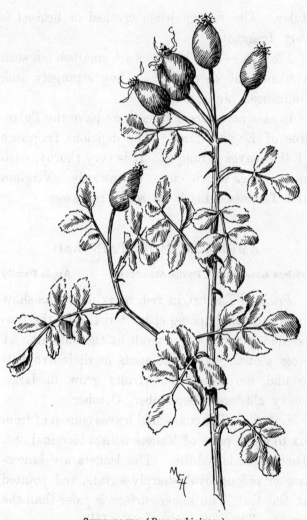

Sweetbrier (*Rosa rubiginosa*)

dular. The foliage when crushed or bruised is very fragrant.

Flowers. — The blossoms are smallish but such a wonderful deep pink. They strangely lack fragrance. June, July.

In this naturalized species, we have the Eglantine of English fame. The delicious fragrance of the leaves is unique. It is very thorny, with stout spines which curve downwards. Virginia and Tennessee mark its southern range.

AMERICAN MOUNTAIN ASH

Sorbus Americana.   Pyrus Americana   **Apple Family**

Fruit. — The bright red, berrylike fruits show externally their pome characteristics by the five-pointed, starlike calyx teeth at the summit. A cross section shows the seeds in their five cells around the core. The fruits grow in large, heavy clusters. September, October.

Leaves. — The compound leaves consist of from six to eight pairs of leaflets with a terminal one. Their stem is reddish. The leaflets are lanceolate or oblong oval, sharply serrate, and pointed at the tip. The under surface is paler than the upper. They are yellow in the fall.

Flowers. — The small white flowers grow in flat compound cymes. May, June.

AMERICAN MOUNTAIN ASH (*Sorbus Americana*)

The Mountain Ash Tree is gorgeous in fruit. The birds, however, do not seem to care for the fruit, neglecting it if other food is available. The American species closely resembles the

European, which is the one usually grown on lawns. Our native tree has a darker bark, smoother leaves and stem, more sharply toothed leaves, and darker, smaller fruit. The tree is more slender in its growth.

The range is from Newfoundland to mountains of North Carolina, west to Michigan and Minnesota.

Sorbus sambucifolia (*Pyrus sambucifolia* of Gray) is much like the preceding but with smaller cymes and larger fruit and flowers. It is a more northern tree, northern New England limiting its southern range. It occurs near Lake Superior and westward.

RED CHOKEBERRY.   DOGBERRY

Aronia arbutifolia.   Pyrus arbutifolia   **Apple Family**

Fruit.—The fruit grows in an erect cymelike cluster. Each pome, small and berrylike though it be, shows its resemblance to an apple in the calyx teeth and the dried stamens which it bears at the apex. A vertical section shows the "core," and a cross section the five cells with their normally two seeds. The flesh is reddish or

RED CHOKEBERRY (*Aronia arbutifolia*)

67

dark in color and not very thick. The separate fruits are reddish, globose or pear-shaped, and about the size of a large huckleberry. They are sweet but rather dry and astringent. They often remain long on the bushes, as birds do not seem to care for them.

Leaves. — The margins of the oblanceolate or oblong leaves have fine rounded teeth. The petioles are short; the apex is obtuse or sharply narrowed; and the base, narrowed. The upper midrib is glandular. The under surface of the leaf is woolly. When the leaves change they assume dark red and orange shades.

Flowers. — The white, rose-shaped flowers grow in compound downy corymbs.

The chokeberry is a shrub from one to three feet high, occasionally reaching a height of twelve feet. It is largest in swamps and moist thickets but often grows in dry places. It is common from Nova Scotia south, and westward to Minnesota.

JUNEBERRY (*Amelanchier Canadensis*)

70

SERVICE BERRY. JUNEBERRY. MAY CHERRY

Amelanchier Canadensis   **Apple Family**

Fruit. — The berrylike pomes vary in color from a red to an almost violet-blue. They are covered with a slight bloom. The calyx lobes, at the summit, inclose several dried filaments. The ovary is five-celled with two ovules to a cell, but as the fruit develops a false partition grows between the two ovules of each cell, making the fruit ten-celled with one seed in each, if all the ovules develop. The fruits are generally globose, and grow in racemes on rather long, slender stems. They are sweet and delicious in flavor. They ripen in June ; hence the name of Juneberry.

Leaves. — The ovate or ovate-oblong leaves are sharply toothed, rounded or heart-shaped at the base, and acute at the tip. When young they are hairy, but become smooth.

Flowers. — The white flowers, with their strap-shaped petals, grow in loose, drooping racemes at the ends of branchlets.

This species is a tree from ten to thirty feet in

height. It is said to fruit sparingly and to be soon robbed of its fruit by the birds, — bluebirds, robins, cedar birds, orioles, downy and hairy woodpeckers. I have been fortunate enough to know it in prolific seasons, when the trees stood laden with red and purplish fruits for two or three weeks.

Amelanchier is a plant which is much influenced by climatic conditions. Two apparently different types exist east and west of the Rockies. On the Rocky Mountains the two merge into each other until they cannot be distinguished.

The fresh and dried fruits of one variety are said to have been used by the Indians. Dr. Hooker says they make a pudding which is nearly equal to plum pudding.

Amelanchier Botryapium, or Shad Bush, is a lower plant, sometimes a shrub. The young leaves are more woolly, the racemes shorter and thicker, and the fruit smaller, on shorter stems, and more juicy. It grows in low wet spots or in swampy woods.

COCKSPUR THORN

Cratægus Crus-Galli   **Apple Family**

Fruit. — The external appearance is decidedly that of a pome, with its five, persistent, sharply pointed calyx lobes at the summit. The seeds are, however, bony, like those of a drupe. The fruit is red and nearly globular. September and throughout the winter.

Leaves. — The leaves are inversely egg-shaped, with pointed or rounded apex. The leaf tapers toward the base, the margin of which is entire. The remainder of the margin is toothed. The upper surface is smooth and shining and the lower one is paler. Yellow and red are the colors of the fall foliage.

Flowers. — The fragrant white flowers grow in irregular corymbs.

This Thorn becomes a small tree. It has long, slender, sharp thorns. It is not very common as a native, but is well adapted for cultivation.

LARGE–FRUITED THORN. DOTTED–FRUITED THORN

Cratægus punctata   **Apple Family**

Fruit. — The red or yellow globular pomes
are dotted with whitish dots. They grow on
dotted hairy peduncles in leafy corymbs. The
bony nutlets are rounded and somewhat grooved.
The flesh is dry and tough but rather pleasant
flavored. The calyx lobes crown the summit.
The fruits are abundant. September.

Leaves. — The inversely egg-shaped leaves
are acute or rounded at the apex and taper
toward the base, finally forming winged petioles.
The margin above the middle is serrate. The
veins beneath are prominent and usually hairy.
The leaves are rather thick and firm.

Flowers. — The white flowers grow in somewhat leafy clusters. The flower stems are
downy.

This is a thick spreading tree with horizontal
branches. It is not very tall. It frequently
grows in thickets. The bark is rough. The
thorns are sharp and light brown. The Duke of

Argyle is said to have introduced this tree into English gardens.

The seeds of the Thorn fruits, or haws, are so hard that it requires a considerable time for their germination. In some parts of France when a hawthorn hedge is wanted, the haws are fed to turkeys. The seeds are uninjured by the digestive process but the hard coats are somewhat softened, and germination is more readily secured. It extends along the Alleghanies into Georgia and Alabama; Quebec and Ontario are its northern limits.

SCARLET THORN

Cratægus coccinea   **Apple Family**

Fruit. — The globose or ovoid pomes grow in small clusters, two or three fruits in each. They are bright red, on slender stems, and bear calyx lobes at the top. The flesh is thin. The three or four nutlets are deeply ridged along the back. The fruit is rather sweet and dry. September, October.

Leaves. — The broad-ovate leaves grow alternately on slender stems which are grooved above.

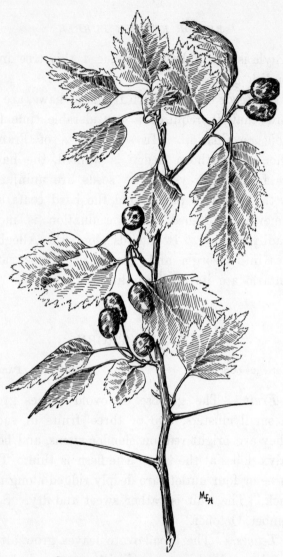

Scarlet Thorn (*Cratægus coccinea*)

76

The leaves are finely toothed and deeply cut, almost lobed along the upper half. The outline somewhat resembles that of a White Birch leaf. The under surface is paler than the upper. Yellow is the fall color.

Flowers. — The rather large white flowers grow on slender stems, in clusters. They have a strong, disagreeable odor. May.

This is a low tree with crooked, spreading branches, ashy gray or light brown bark, and stout thorns attaining maturity on third-year growths. The plants like moist soil but will grow in pasture lands, where they form thickets, the thorns protecting them from destruction by grazing animals. In New England this thorn is generally larger than the other species.

Cratægus macracantha has longer thorns, thicker leaves, stouter stems, and larger flowers and fruits. The leaves are sometimes doubly serrate.

Cratægus mollis varies chiefly from *Cratægus coccinea* in having hairy leaves, twigs, and leaf stems. It is about two weeks earlier.

Partridges are fond of the Thorn fruits, and in the good old days, when snaring the birds was not "prohibited by law," the bright little apples were used for bait. "When I was a boy," said an elderly man to me only the other day, "we used to dig narrow paths in the snow; set up two sticks with a string stretched across them, and a loop of horsehair hanging from the string; scatter Thorn-apples along the path; and await results. Many a plump bird have we found the next morning, unable to free himself from the horsehair loop, through which he endeavored to reach the edible fruits beyond."

PEAR THORN

Cratægus tomentosa   **Apple Family**

Fruit. — The pear-shaped, seldom round, drupe-like pome is red or orange-red. It is crowned by the erect calyx lobes. The flesh is thin and the seeds are bony. They are rounded, and have on the back two faint grooves. September, October, and persistent.

Leaves. — The leaves are firm and leathery, and are borne on petioles which are margined to the base by the tapering leaves. The margin is doubly serrate, and sometimes so deeply cut near the apex as to form lobes. The under surface is downy along the veins.

Flowers. — The ill-scented white flowers grow in leafy corymbs on downy flower stems. The calyx lobes are likewise covered with down.

This small tree has dark brown to gray bark and sharp axillary thorns. It is quite widely distributed throughout the country, but is not so common in the Northern states. Central New York contains flourishing growths of it. The fruits cling to the tree until spring.

WILD YELLOW OR RED PLUM

Prunus Americana   **Plum Family**

Fruit. — The fleshy drupe is yellow or reddish, somewhat whitened with a bloom. It is globose, with a slight depression at the tip. It grows laterally on a stout, rather short, stem. The skin is thick and tough, the flesh quite thick, and the stone rather smooth, with quite sharp edges. August, September.

Leaves. — The ovate leaves are coarsely or doubly serrate. They are nearly smooth, or somewhat hairy along the veins on the lower surface. The apex terminates in a long tip.

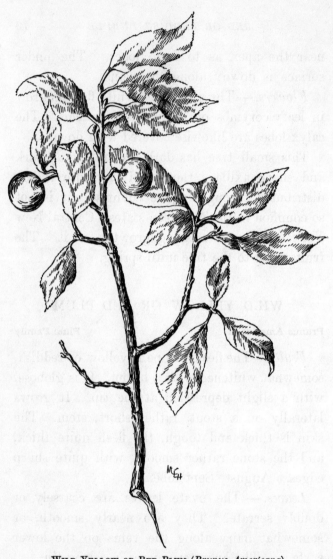

WILD YELLOW OR RED PLUM (*Prunus Americana*)

Flowers. — The white flowers precede the leaves, and grow in lateral sessile umbels. April, May.

Authorities differ much as to its range. I have known the tree along Connecticut roadsides, especially in the eastern portion of the state. The tree is small and thorny, and is quite prolific.

CANADA PLUM. HORSE PLUM

Prunus nigra   **Plum Family**

Fruit. — This plum differs from the preceding in being oblong-oval. It is from an inch to one and a half inches long. The red or orange-colored skin is tough, and the flesh clings to the flat stone. It is of pleasant flavor. August.

Leaves. — The ovate or obovate leaves are not so sharply serrate as those of *Prunus Americana*, nor bristle-tipped. The apex is long-pointed and the base wedge-shaped or somewhat heart-shaped.

Flowers. — The white flowers are larger than in the preceding species, and change to pink after opening. They grow in two- to three-flowered umbels. May.

This is a species of more northern range than
Prunus Americana. In Canada the fruits are
extensively marketed, being used raw or for
preserving. The plant occurs in northern New
England, but has not been reported in Connecticut, and is but occasional in Massachusetts.
It flourishes in the St. Lawrence valley and as
far west as Lake Manitoba. It follows streams,
grows along fences, and springs up in thickets.

BEACH PLUM

Prunus maritima   **Plum Family**

Fruit. — The purplish or red bloom-covered
drupe is globular and from one-half to one
inch in diameter. It hangs by a slender stem.
The stone is thin and sharp on one edge and
rounded on the other. It is usually pointed at
each end. August, September.

Leaves. — The ovate or oval leaf has a rounded
base and an acute apex. It is finely serrate.
The leaves are arranged alternately. They often
have one or two glands at the base. Dark red
and orange are the autumnal colors.

Flowers. — The white flowers grow profusely
in umbels along the sides of the branches.

They open before the leaves appear. April, May.

This rather low shrub is a *habitant* of sandy or stony beaches, and sometimes grows in waste places twenty miles or so inland. It grows in clumps and often fruits abundantly. The plums are sweet when ripe, and in some places are gathered and sold for preserving.

WILD RED CHERRY. BIRD CHERRY. PIN OR PIGEON CHERRY

Prunus Pennsylvanica   **Plum Family**

Fruit. — The small light red drupes grow in clusters of from two to five. These clusters grow from the leaf axils or take the place of leaves at the end of the previous year's shoots. They often occupy a leafless space of six or more inches along the branches, with leaves above and below them. The slender fruit stem is from three-quarters to an inch in length. Each cherry is globular, about the size of a pea, and retains at the tip a remnant of the style. The flesh is thin and sour. The stone is large in comparison with the whole

WILD RED CHERRY (*Prunus Pennsylvanica*)

84

fruit, is nearly globular, and has noticeable grooves and ridges along one side.   July.

Leaves. — The leaves are oblong-lanceolate, with pointed apex and rounded base.   They are finely serrate, and in arrangement are alternate or in pairs.   They are a bright shining green above and lighter beneath.   In autumn they change to a bright yellow.   The petioles are slender and grooved.

Flowers. — The white cherrylike flowers grow in umbels of from five to eight blossoms

The Wild Red Cherry is a small tree from twenty to thirty feet high.   It is especially a tree of the Northern forests, but extends southwards along the mountains, attaining its greatest size in the mountains of Tennessee.   It often springs up abundantly over cleared lands and is found along ravines.

George Emerson tells of using the dry beds of hill streams as a footpath and of finding there numerous stones of the Wild Red Cherry, although there were no trees of the kind within a considerable distance.   Water, as well as birds, seems in this case to act in scattering the seeds.

The bark of the tree is reddish brown with raised, rusty-looking dots, and has the •common cherry characteristic of peeling off in horizontal strips.

CHOKE CHERRY

Prunus Virginiana   **Plum Family**

Fruit. — The drupes, which are about the size of peas, grow in long drooping clusters at the ends of leafy branches of the season's growth. Each cherry is borne on a short stem nearly equal to it in length. It is globular or oval, with a thin, shiny, dark red or nearly black skin. Yellow fruits have been found. The pulp is yellow, juicy, and rather sweet. The cherries vary much in flavor, but in all cases are more or less astringent. July, August.

Leaves. — The oval or obovate leaves grow from rounded stems which are grooved on the upper surface. Two or four glands are borne on the margins of these grooves. The leaves are rounded or wedge-shaped at the base and sharply pointed at the apex. The margins are sharply serrate. The upper surface of the leaf is bright green and the lower one is lighter.

Flowers. — The small, white, cherrylike flowers grow in loosely flowered, erect, or spreading racemes. The petals are more rounded than those of the Wild Black Cherry. April, May.

This shrub sometimes becomes a small tree. The largest growths are found in Nebraska, Indian Territory, and Texas. The trunk rarely has a diameter of more than two or three inches.

The plant is decorative in fruit, with its clusters of shining, jewel-like spheres. The fruit of some shrubs is quite pleasant to the taste, while one cherry from another will "pucker" lips, tongue, and roof of mouth, and set one's teeth on edge. The skin seems to possess more of the astringent quality than the flesh.

Bluebirds, robins, cedar birds, crows, kingbirds, hairy woodpeckers, and flickers are fond of the fruit. Bears are said to aid in scattering the seed. As for children, how they will fur their tongues with bunch after bunch of the cherries! It is almost impossible to remove the stain of this fruit from clothing.

The Choke Cherry has an extended range from within the Arctic Circle to the Gulf of Mexico and across the continent. It is a familiar feature of roadside and fence-row growth and often grows near streams.

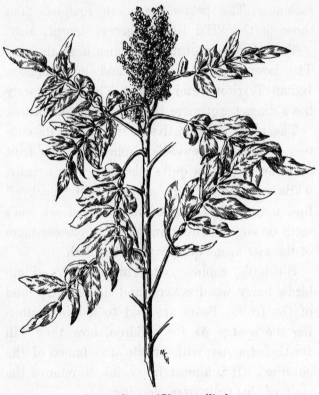

DWARF SUMAC (*Rhus copallina*)

DWARF SUMAC

Rhus copallina   **Sumac Family**

Fruit.— The panicle of bright red fruit is
quite open. Each drupe is compressed and

rather short, bearing the stigmas at the top. Gray dots are scattered over the fruits. The berry is acid. Persistent.

Leaves. — There are from nine to twenty-one leaflets, with noticeable wings along either side of the stem between them. This is a distinguishing feature of the species. The leaflets are often entire and shine above as if polished. The under surface is lighter and downy. In autumn the leaves become a rich purple.

Flowers. — The fertile and sterile flowers are in separate clusters, the pistillate in much smaller ones than the staminate.

This sumac, like *Rhus hirta*, is pubescent, but may be readily distinguished by its winged, seemingly jointed, petioles. The term "Dwarf" is somewhat misleading, as the plant sometimes reaches a height of eighteen or twenty feet. It is a beautiful shrub, growing on rocky hills.

STAGHORN SUMAC

Rhùs hirta. Rhus typhina   **Sumac Family**

Fruit. — The small dry drupes are borne in a terminal, compound, compact cluster. Each fruit is one-seeded, has a very thin coat, and is thickly

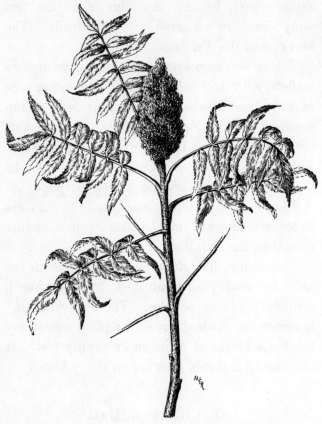

STAGHORN SUMAC (*Rhus hirta*)

covered with silky hairs. The calyx persists at
the base. The fruits are acid. August, and
persistent through the winter.

Leaves. — The leaves are compound and odd-pinnate, with from eleven to thirty-one leaflets. The petioles are red above and green below and densely covered with hairs. The leaves are alternate. The leaflets are oblong-lanceolate, sessile, sharply serrate, acute at the tip and rounded at the base. They are paler beneath and hairy. The autumnal colors are brilliant; red, yellow, and orange.

Flowers. — The sterile and fertile flower panicles are usually on different plants, although they are occasionally on the same one. They are greenish yellow. June, July.

Rhus hirta is sometimes called Velvet Sumac, and appropriately so, for branches and stalks are so densely coated with soft hairs as to resemble, both to the sight and touch, a velvet covering. This hairy appearance, together with the irregularly forked branches, somewhat resembling the horns of a young stag, has given rise to its other popular name, Staghorn Sumac.

It sometimes reaches the stature of a small tree. The brilliancy of its autumnal foliage is a great addition to the hills which it frequents. Sometimes a whole pasture is aglow with it. For two successive springs I have seen my first robin on the sumac bushes, dining on the fruits which have been preserved for it through the winter. The catbird includes sumac drupes in his spring diet. The taste of the berries after their exposure to the cold of winter is much less acid than in the fall. The bark and leaves, because of their astringent qualities, are useful in tanning.

SMOOTH SUMAC

Rhus glabra   **Sumac Family**

Fruit. — The dry drupes grow in a more open, compound cluster than those of *Rhus hirta*. The smaller clusters composing the fruit panicle alternate in much the same fashion as the leaves. The calyx persists at the base of each drupe, which is covered with fine red hairs. The fruit is rounded and flattened on two sides. September, and persistent.

Leaves. — The compound pinnate leaves, with terminal leaflets, grow on smooth, reddish petioles. Authorities differ as to the number of the leaflets. They are oblong-lanceolate, sessile, toothed, and have a long point at the apex and rounded base. They are whitened beneath and smooth. The foliage is gorgeous in crimsons and gold in the fall.

Flowers.— The greenish flowers grow in terminal, much-branched heads. June, July.

This is a smooth sumac which does not attain the size often reached by its velvet-coated brother. The two sumacs frequently grow together and form clumps. Their deep roots render them difficult of extermination. The berries are sometimes used in dyeing reds.

FRAGRANT OR SWEET–SCENTED SUMAC

Rhus aromatica. Rhus Canadensis   **Sumac Family**

Fruit. — The globose, red, downy drupes are in short clustered spikes.

Leaves. — The compound leaf is composed of a terminal, short-stalked leaflet and two lateral sessile ones. The terminal one is sometimes three-cleft. The bruised leaves are rather fragrant. In autumn the leaves are orange and red.

Flowers. — The yellowish green blossoms appear before the leaves and are borne in short spikes.

This is a low, straggling shrub, growing in patches on sandy or rocky banks. It occurs in western Vermont and thence west to Minnesota. It is not poisonous.

AMERICAN HOLLY

Ilex opaca   **Holly Family**

Fruit. — The globular red drupes are borne on short stalks along the recent growths or from the leaf axils, looking like big red-headed pins partly stuck into the branches. The remnant of the stigma at the summit appears as a black spot. The usually four-parted calyx lobes are at the base. Each drupe contains four to six small nutlets, which are ribbed, veiny, or one-grooved on the back. They are somewhat triangular in shape. The flesh is yellow and rather thin. Persistent.

Leaves. — The thick, leathery, evergreen leaves are shining above and paler beneath. They have large teeth which terminate in spines. They are oval in outline, with pointed apex and pointed or angular base.

Flowers. — These are usually diœcious. The small white or greenish blossoms appear in June.

The sterile or partly sterile ones grow in clusters, usually in the axils. The fertile ones are solitary.

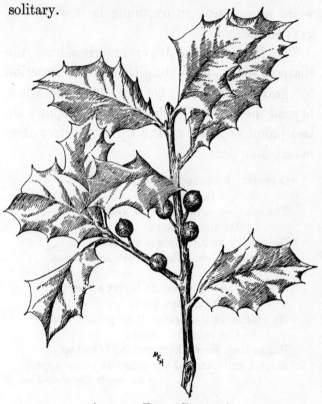

AMERICAN HOLLY (*Ilex opaca*)

This is a small tree with light gray bark, lighter than that of the beech, which it somewhat resembles. Its evergreen leaves and bright

persistent berries make the fertile tree very ornamental. In its native wilds, it often presents a
weird appearance, so overhung is it with soft,
grayish lichens.

The leaves of the lower branches have the
sharpest spines, preventing the tree's destruction
by grazing animals. On the upper branches,
beyond the reach of such enemies, the spines are
less prominent, and at the tip of the tree they
nearly disappear.

> " O reader! hast thou ever stood to see
>   The holly-tree?
> The eye that contemplates it well perceives
>   Its glossy leaves
> Ordered by an intelligence so wise
> As might confound the atheist's sophistries.

> " Below, a circling fence, its leaves are seen
>   Wrinkled and keen;
> No grazing cattle through their prickly round
>   Can reach to wound;
> But as they grow where nothing is to fear,
> Smooth and unarmed the pointless leaves appear."
>   —SOUTHEY'S *The Holly-Tree.*

The glossy leaves and showy berries have long
been associated with the Christmas season. The
wood is hard and capable of a beautiful polish.
It is used for cabinet making, whip handles,

engraving, etc. Our species closely resembles the European Holly, differing from it in having less glossy leaves and duller fruit. Holly occurs more or less frequently in New England and New York. It is abundant from New Jersey along the coast to the south, and in the Gulf States. Holly is dependent upon sea air, and will not grow much more than a hundred miles inland.

Ilex monticola, or Large-leaved Holly, grows in the Catskills and along the Alleghanies to Alabama. It is usually a shrub, rarely becoming a tree. It bears a reddish drupe containing ribbed nutlets. The leaves are thin, deciduous, ovate, and sharply toothed. The fertile flowers grow on very short stems and are solitary. The sterile ones are clustered.

BLACK ALDER. VIRGINIA WINTER BERRY

(*For illustration, see Frontispiece.*)

Ilex verticillata　　　　　　　　　　**Holly Family**

Fruit. — The bright, scarlet, glossy drupes are about a quarter of an inch in diameter. The dark stigma is at the top and the persistent calyx is at the base. The pulp is yellowish, and

contains three to eight lunate smooth nutlets.
The fruits grow on short stems and are solitary
or in clusters. They appear as if arranged
spirally around the branches. The flicker is
said to eat the berries. September, and clinging
long after the leaves fall.

Leaves. — The leaves turn black in autumn.
They are oval or wedge-lanceolate, acute at the
apex, toothed, smooth above and hairy below,
along the depressed veins.

Flowers. — The small, polygamo-diœcious
flowers are solitary or clustered in the axils.
May, June.

The fence and stone-wall growth is brightened
in the fall by the Black Alder with its scarlet
berries. These are said to be eaten by flickers,
and its growth along fence rows would suggest
its dispersal by birds. The bushes with the
berries snow-laden are a beautiful sight. I was
glad to recognize these bright wild fruits in
the windows of New York City florists,
placed amidst fantastic orchids and customary
Christmas decorations. The plant ranges
throughout the eastern part of the United
States as far west as Missouri. It also occurs
in Nova Scotia.

SMOOTH WINTER BERRY

Ilex lævigata   **Holly Family**

Fruit. — The rich orange-red drupes are larger than the preceding and ripen earlier. They grow on peduncles in length equaling their diameter. September.

Leaves. — The thin, light green, oval or oblong leaves have a glossy luster on either side. The apex is acute and often has a twisted point; the base is also acute. The leaves are obscurely toothed. They are bright yellow in the fall.

Flowers. — The small white flowers are perfect or diœcious, and grow in the leaf axils on slender stems.

The extent of this species is from Maine to the mountains of Virginia. Its range is much more limited than that of the preceding plant. Its yellow autumnal coloring is one distinguishing feature.

WILD OR MOUNTAIN HOLLY

Ilicioides mucronata.   Nemopanthes fascicularis
Holly Family

Fruit. — The pale crimson, nearly globular, berrylike drupe grows from the leaf axil, on a red stalk, an inch or more in length. The flesh is yellowish and incloses four or five faintly ribbed stony nutlets. September.

Leaves. — The oblong deciduous leaves grow on slender stems. They are entire or faintly toothed and acute or bristle-tipped at the apex.

Flowers. — The flowers are small, white, and polygamo-diœcious. May, June.

The long, threadlike peduncles are distinctive features of this much-branched shrub. It has an ash-gray bark. Its *habitat* is in damp woods along the mountains in Virginia, and northwards. It is found west to Indiana and Wisconsin.

STRAWBERRY BUSH

Euonymus Americanus   **Staff-tree Family**

Fruit. — The rough, warty, crimson capsule opens its usually five pods and discloses the scarlet arils of the seeds. There are one to four seeds in each cell.

Leaves. — The ovate to oblong-lanceolate, nearly sessile, leaves are bright green with a pointed apex.

Flowers. — The small flowers grow in loose cymes from the leaf axils. June.

This is an erect shrub, sometimes six feet high. It grows along the wooded banks of streams from New York and Illinois, southward.

Euonymus obovatus, Running Strawberry Bush, is low and straggling. The leaves are inverse egg-shaped, and grow on short stems. The flowers are smaller and earlier than in the preceding species. The fruit is usually three-celled. It has a more limited range than Strawberry Bush, its southern boundaries being Pennsylvania, Indiana, and Kentucky.

BURNING BUSH. WAHOO. SPINDLE TREE

Euonymus atropurpureus   **Staff-tree Family**

Fruit. — The smooth fleshy pod or capsule is three- or four-lobed and purple in color. The pods open, when mature, enough to disclose the

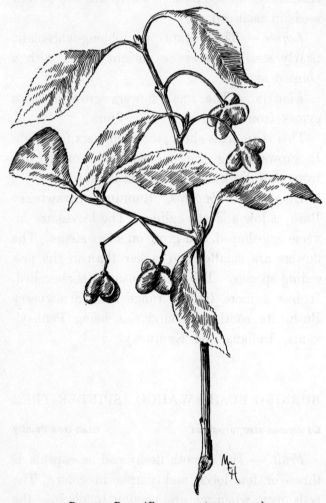

Burning Bush (*Euonymus atropurpureus*)

102

bright red ariled seed. The fruits grow on long, drooping stems and hang late on the branches. The fruit is said to be poisonous. October.

Leaves. — The thin leaves are ovate or elliptical, pointed at the apex and pointed or blunt at base. They are finely toothed.

Flowers. — The dark purple flowers grow in few-flowered clusters on drooping stems.

In New England this plant appears only as a cultivated shrub. In New York, west to Wisconsin and Nebraska, and southward, it is found along the wood borders. In Arkansas and Indian Territory it reaches tree size.

WAXWORK. SHRUBBY OR CLIMBING BITTERSWEET

Celastrus scandens   **Staff-tree Family**

Fruit. — The yellow or orange berrylike capsule opens and bends backward its two to three valves, disclosing the scarlet arils which surround the seeds. There are three cells, with one or two brownish oblong seeds in each. The fruits grow in a loose, spikelike cluster. September.

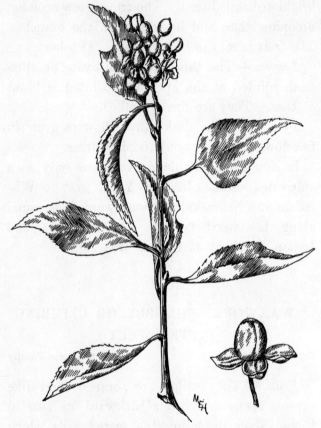

Shrubby or Climbing Bittersweet (*Celastrus scandens*)

Leaves. — The ovate-oblong leaves are usually pointed at the apex and at the base. They are slightly toothed and alternate in arrangement.

Flowers. — The staminate and pistillate flowers often grow on different plants. They form long, loose spikes. June.

The fruit of this plant is highly decorative, and if gathered before the capsule opens will develop in the house and remain in good condition throughout the winter. The woody vine coils upon itself, and climbs over fences and trees. I shall never forget the glory which a roadside nook revealed one bright autumnal day. The dark Pine and White Birch were growing together, and winding in and out and over both gleamed the bright berries of the Bittersweet. It was too beautiful to spoil, and we left it undisturbed. It grows from North Carolina northward, but is said to be rare in the White Mountain country.

LEATHERWOOD. MOOSEWOOD

Dirca palustris   **Mezereon Family**

Fruit. — The oval, shining, reddish drupes are solitary or from two to three in a cluster. Each fruit contains a compressed ovate seed. The flesh is thin and tough. The fruit matures rapidly and falls early.

Leaves. — The leaves are oval or inverse egg-shaped. The under surface is much lighter than the other. The petiole is short.

Flowers. — The light yellow flowers appear before the leaves. Usually three come out of the same bud, with their short stalks cohering. April.

It is from the toughness of its bark that this shrub receives its name, Leatherwood. The wood is quite brittle, but it is almost impossible to break the bark. The Indians knew of this quality and utilized it for thongs. The twigs are used in basket making with good effect. The plant grows in moist places in woods from New Brunswick to Minnesota, and south.

CANADIAN BUFFALO BERRY

Lepargyræa Canadensis.   Shepherdia Canadensis
Oleaster Family

Fruit. — The fruit externally resembles a berry. The fleshy, four-cleft calyx, however, incloses a smooth nut or an achene, making the accessory fruit drupelike. It is yellowish red, oval, small, and insipid. July, August.

Leaves. — On short, hairy stems are borne the ovate or oval opposite leaves. These are entire, obtuse at apex, narrowed toward the base, densely silvery scurfy beneath, and smoother and greener above. The scurf is often brownish.

Flowers. — The small yellow flowers are dioecious. The pistillate have the ovaries inclosed in a four-parted, urn-shaped calyx tube, closed at the mouth by an eight-lobed disk.

A low, thornless shrub is the Canadian Buffalo Berry. It has scurfy young shoots. It likes rocky banks. It is a northern plant, extending down into Vermont and New York.

GINSENG

Panax quinquefolium.   Aralia quinquefolia
Ginseng Family

Fruit. — The fruit grows in a simple umbel. The berries are bright red, and sometimes in joined pairs. They are somewhat flattened, drupelike, and have two or three seeds.

Leaves. — The compound leaves grow in a whorl of three. Each leaf has five leaflets, — seldom more, — and its appearance is somewhat

like that of a Horse-chestnut leaf. Each leaflet
is ovate or obovate, thin, and sharply toothed,
with a pointed apex and narrowed or rounded
base.

Flowers. — The greenish yellow, polygamous
flowers grow in small, simple umbels. July,
August.

The root of the Ginseng is in such demand
for its supposed medicinal value that the plant
has become quite rare. Recently Ginseng plantations have been started to supply the demand
for the root. The Chinese, especially, prize it as
a remedy for fatigue and a preventive against old
age. The Chinese name for it is *Jinchen*, meaning manlike, from its fancied two-legged shape.
Its range is south to Alabama and west to Minnesota, Nebraska, and Missouri.

LOW OR DWARF CORNEL. BUNCHBERRY

Cornus Canadensis   **Dogwood Family**

Fruit. — The bright red drupes grow in a compact bunch at the summit of the stem. They
are globose and bear the calyx teeth at the tip.
The solitary stone is smooth and nearly globular.

BUNCHBERRY (*Cornus Canadensis*)

Leaves. — The upper leaves are nearly stem-less, in a whorl of four or six at the top of the stem. One or two pairs of scalelike leaves sometimes occur along the stem. The leaves are entire, acute at each end, and ovate or oval.

Flowers. — The small greenish flowers are in a close cluster, and surrounded by four white bracts. May–July.

The Bunchberry is reported as growing profusely among the White Mountains and the Adirondacks. It is very attractive in fruit. "But," said a woman who was exclaiming over them, "the people who live among them all the time don't even know their names and hardly notice them." Truly, many there are who, having eyes, see not the beauties of their common environment.

New Jersey, Indiana, and Minnesota are the limits of southern range. It extends far northward and westward.

FLOWERING DOGWOOD

Cornus florida   **Dogwood Family**

Fruit. — The small ovoid drupes are bright red and grow in small bunches. They are ovoid and bear at the tip the calyx and the remnant of the style. The flesh is bitter and

FLOWERING DOGWOOD (*Cornus florida*)

111

unpleasant. The stone is smooth and channeled. September.

Leaves. — The leaves are oval with a prolonged apex. They are narrowed at the base
and entire. The upper surface is shining and
the lower one lighter and often downy. The
autumnal colorings are rich in scarlets and
crimsons.

Flowers. — The inconspicuous greenish flowers
grow in heads, surrounded by a showy white
involucre of four parts, often mistaken for the
petals.

This shrub or small tree grows readily in the
shade of other trees. It is showy in springtime,
with its large white bracts surrounding the
flower clusters and acting as signals to the insects that assist in the fertilization of the inconspicuous blossoms. These bracts are, in reality,
developed bud scales, which are not in this
plant thrown off when their protective offices
against the cold and storms of winter have been
performed. The blossom is the "corn sign"
of the New England farmer.

In the fall, the red fruit clusters amidst the autumnal foliage present a fine showing. The fruit
lingers throughout the fall, and after the frosts

have somewhat changed its taste is eaten by robins. The bitter bark is somewhat similar in its action to Peruvian bark and is sometimes substituted for it.

The plant grows in dry woods from southern New England west to Ontario and Minnesota and south to Florida and Texas.

SPRING OR CREEPING WINTERGREEN CHECKERBERRY. BOXBERRY TEABERRY

Gaultheria procumbens   **Heath Family**

Fruit. — The actual fruit capsule is five-celled, with many seeds in each cell. It is like a flattened sphere in shape, and its flesh is very thin. This capsule, however, is nearly inclosed in a thickened, fleshy, red calyx, which gives to the whole the appearance of a berry. The developed calyx plainly shows its five lobes. It is subtended at the base by two small bracts.

The so-called berries grow on short, drooping stems from the leaf axils. They are usually solitary. The berry is dry and mealy, but has a delightful aromatic flavor similar to the sweet birch.

SPRING OR CREEPING WINTERGREEN (*Gaultheria procumbens*)

Leaves. — The usually few, thick, evergreen leaves are borne at the ends of the branches. They are alternate, ovate and glossy above with a whitened under surface. They are sparsely toothed with bristle-like teeth. The petioles are short and reddish. The young leaves are tender and delicious in flavor.

Flowers. — The usually solitary, white, nodding flowers are waxy and vaselike.

When we search woods and moist banks in June for the " Young Wintergreen " we are apt to find last year's berries still lingering. The new fruits ripen in the fall, and serve during the winter as food for the birds. This plant is one example of red fruits contrasted with evergreen leaves. The branches grow from a creeping or underground stem. The plant is found in various localities. Its range is southward from Maine and west to Michigan.

RED BEARBERRY

Arctostaphylos Uva-Ursi   **Heath Family**

Fruit. — The drupes grow in short clusters and retain the calyx at the base of each fruit. They are red, and the flesh is mealy and tasteless. The five nutlets become inseparably united. Each shows a line along its back. The fruits remain on the plant during the year.

Leaves. — The thick evergreen leaves are inversely egg-shaped. The apex is obtuse and the base narrows to a short, downy stem. The upper leaf surface is shining and the lower one paler; both are smooth. The margin is entire or hairy. The leaves are somewhat crowded towards the ends of the branches. In winter, the upper surface becomes somewhat brown and the under one reddish.

Flowers. — The drooping, white or pink, pitcher-shaped flowers grow in a short end cluster. The stigma matures from two to five hours before the anthers shed their pollen. The opening of the flower is bearded or filled with a "woolly thicket" to keep out winged insects.

This evergreen shrub trails over rocks and sandy wastes. It abounds in the Alps and in other mountainous sections of Europe, as well as in the northern countries of Europe and Asia. It prevails throughout Canada and south to New Jersey, Pennsylvania, Illinois, Michigan, Nebraska, Colorado, and California.

Its nomenclature is varied, the plant being known as Foxberry, Bear's Grape, Mealberry, Barren Myrtle, etc. The fruits serve as food for grouse and partridges. The plant is used in tanning, especially in parts of Europe, and is also used for dyeing. The Indians smoke the leaves as a preventive against malarial disorders. It is known among them as Kinnikinic.

COWBERRY. MOUNTAIN CRANBERRY
FOXBERRY

Vaccinium Vitis-Idæa   **Huckleberry Family**

Fruit. — The four- to five-celled, many-seeded berry is dark red, acid, and often bitter. It is less than half an inch in diameter. The fruits grow in short terminal clusters. August, September, and persistent.

Leaves. — The evergreen leaves are thick and leathery, with somewhat shining upper faces and paler under ones that are dotted with bristly black points. They are somewhat similar to box leaves but darker. They are obovate or oval and short-stemmed.

Flowers. — The nodding, white or pink, bell-shaped flowers are in short terminal clusters. June.

This is essentially a northern plant, extending far to the north, and appearing in our range in the mountains and along the coast of New England and west to the northern shore of Lake Superior. It also occurs in Europe and Asia. In northern Europe it flourishes profusely, and is there used in making a jelly which is served with roast beef and deer flesh. It is also used for colds and sore throats. The flavor of the fruit seems to improve towards the north, much of the bitterness being lost. Birds feed upon large quantities of the berries during their migrations. Bears, too, are fond of them. They uproot the bushes to get the hidden fruit near the ground. The shrub is low, only about a foot high, with the upright branches growing from creeping stems.

SMALL OR EUROPEAN CRANBERRY

Oxycoccus Oxycoccus.   Vaccinium Oxycoccus
Huckleberry Family

Fruit. — The globose berry is red and when young is often spotted. It is rather smaller than the American Cranberry, is acid, and not often marketed. The berry is four- or five-celled and many-seeded. August, September.

Leaves. — The small, thick, evergreen leaves are whitened beneath. They are ovate and entire. The margins are rolled backwards. The apex is pointed and the base rounded or heart-shaped.

Flowers. — The pale rose-colored, nodding blossoms have the corolla nearly divided into four or five parts. The anthers converge to form a cone. May–July.

The ascending branches rise to a height of from six inches to a foot and a half from a creeping stem which sends out roots at the nodes. Patches of cranberries are thus formed, usually in peat bogs. They grow as far south as New Jersey and west to Michigan. In Canada they extend from Labrador to Alaska and British Columbia. They also grow in Europe and Asia.

LARGE OR AMERICAN CRANBERRY

Oxycoccus macrocarpus.   Vaccinium macrocarpon
Huckleberry Family

Fruit. — The berry varies in shape; nearly globular, ovate, or oblong. It grows from the sides of the branches. It is larger than the European Cranberry, and is the species which has been cultivated. It is red when ripe, acid, four- or five-celled, and several-seeded. September, October.

Leaves. — The leaves are similar to those of the preceding species, but are oblong and obtuse at the apex.

Flowers. — The nodding pink flowers grow in clusters. June–August.

This variety is larger and stronger than the preceding and, like it, grows in peat bogs. It grows throughout the north and in the states as far south as North Carolina and west to Minnesota.

It was first cultivated in Cape Cod, which region still holds the highest reputation as a cranberry section. Cranberry plantations have been also established in New Jersey and Wisconsin.

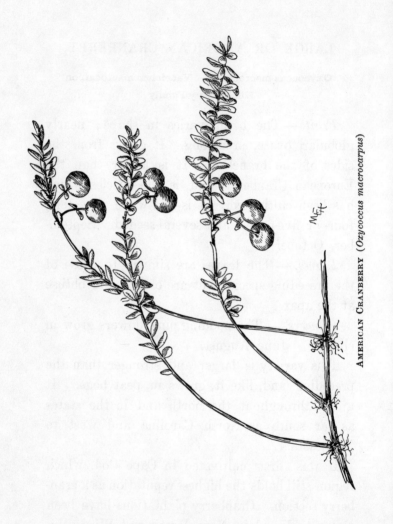

AMERICAN CRANBERRY (*Oxycoccus macrocarpus*)

Eight hundred thousand bushels are estimated to represent a year's production of cultivated Cranberries. The wild Cranberries are also marketed.

One plantation employs a thousand pickers, who camp in tents or cabins during the harvesting. Bailey thus describes the pleasures of the workers: " This picking time is a sort of a long and happy picnic — all the happier for being a busy one. The pickers look forward to it from year to year. They are invigorated by the change and the novelty, and they must come near to nature in the sweet and mellow October days. Those of our readers who have cast their lot with hop-pickers, or who have camped in the clearings in blackberry time, or who have joined in the excursions to huckleberry swamps, can know something of the cranberry picker's experiences. Yet I fancy that one must actually pick the cranberries in the drowsy Indian summer to know fully what cranberry picking is like." — *Evolution of our Native Fruits.*

PHILADELPHIA GROUND CHERRY

Physalis Philadelphica   **Potato Family**

Fruit. — Like all the other fruits of this genus the berry is inclosed in the enlarged calyx. When ripe, the berry fills the calyx or even opens it at the mouth. The undeveloped fruit calyx shows its ten angles and is depressed about the stem. The berry is reddish or purple, quite large, and pulpy. It grows on a slender stem from the leaf axil. The numerous seeds are flattened.

Leaves. — The ovate to ovate-lanceolate leaves usually slant toward the base. They are entire or slightly wavy. They are smooth or a trifle hairy above.

Flowers. — The flowers are yellowish brown with purplish centers. July–September.

This annual is nearly smooth and is tall and upright. It ranges from Rhode Island to Georgia and Texas and west to Minnesota and Nebraska.

NIGHTSHADE. BITTERSWEET

Solanum Dulcamara   **Potato Family**

Fruit. — The oval berries grow in clusters from the sides of the stem. In ripening, the berries change from green through yellow and orange to a bright red, often making a brilliant array of colors in the cluster. The berries are translucent with a thin skin, red pulp, and many seeds arranged around an axial placenta. The five-pointed, starlike calyx is at the base of the berry, which is borne on a stem about as long as itself. The general consensus of opinion seems to be that the berry is poisonous, especially if eaten in any quantity. It begins to ripen in July and hangs long upon the vines.

Thoreau, in describing this fruit, says : "The *Solanum Dulcamara* berries are another kind which grow in drooping clusters. I do not know any clusters more graceful and beautiful than these drooping cymes of scented or translucent, cherry-colored, elliptical berries. . . . Yet they are considered poisonous ; not to look at surely. . . . But why should they not be poisonous ? Would it not be bad taste to eat

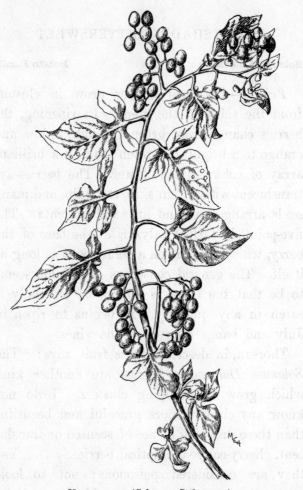

NIGHTSHADE (*Solanum Dulcamara*)

these berries which are ready to feed another sense ?"

Leaves. — The lower leaves are heart-shaped and the upper ones have two lateral lobes at the base. These lobes are sometimes separated from the leaf, forming two lateral leaflets. The leaves are entire and alternate.

Flowers. — The blue, five-parted, wheel-shaped flowers are rendered attractive by the contrast of the blue corolla with the yellow conical group of stamens in the center.

The Nightshade is a climbing vine, sometimes from five to six feet long. My most vivid recollection of it is, as seen from a bridge, growing over a small tree by the side of the river. The tree seemed hung with the graceful, decorative clusters. It is a member of the family which includes such cultivated plants as the potato and egg plant.

It was introduced from Europe. It grows by the side of streams, around houses, and sometimes trails over the stone walls by the roadsides.

MATRIMONY VINE (*Lycium vulgare*)

128

MATRIMONY VINE

Lycium vulgare   **Potato Family**

Fruit. — The oval orange-red berries are solitary or few in the leaf axils. They are small, with the calyx persistent at the base.

Leaves. — The small leaves are oblong or lanceolate and taper into a short stem. The margins are entire.

Flowers. — The purplish or greenish flowers are solitary or two to five in the leaf axils.

In cultivation, this woody shrub is trained over trellises. When growing wild, it trails in masses over any handy support. Its branches are long and drooping. The vine is usually smooth. It often occurs as an escape from cultivation.

PARTRIDGE BERRY

Mitchella repens   **Madder Family**

Fruit.— The scarlet berrylike fruit is really a double drupe, bearing at the summit the teeth of the two flower calices. Each ovary is four-celled with one ovule to a cell, and some fruits have four hard nutlets to each flower ; but often some of the ovules do not develop. The fruit is edible but rather tasteless. The pulp is white and mealy. The berries remain on the vines for a long time, and it is quite common to find flowers, fruit, and even tiny green fruits at the same time.

Leaves. — The round-ovate or heart-shaped, shiny leaves vary from light to dark green. Some have prominent white veinings. They grow in pairs on short stems and are evergreen.

Flowers. — The flowers grow in pairs and are united by their ovaries. They are very dainty with their white linings of soft fine hairs at the throat and an outside coloring of pink. They also have a delicate fragrance.

This vine and its near relative, the Quaker Ladies, are our northern representatives of the family which includes such tropical plants as coffee and cinchona, the latter yielding quinine. *Mitchella repens*, besides belonging to our range, grows in the forests of Mexico and Japan. It frequents dry woods, especially pine forests, and trails its vines in masses around the foot of trees, the base of rocks, and over many a pine needle carpeted space. The contrast of the green vine

PARTRIDGE BERRY (*Mitchella repens*)

131

and its bright berries with the brown of the needles is so charming that one wonders that it has not been copied for our indoor carpetings. A low glass dish filled with wood earth and containing a root or two of Ebony Fern, a little Rattlesnake Plantain, and a few vines of the Partridge Berry will serve all winter to shut-ins as a most delightful reminder of the woods.

The plant is named for Dr. John Mitchell, an early Virginian botanist.

RED–BERRIED ELDER

Sambucus pubens.   Sambucus racemosa
Honeysuckle Family

Fruit. — The red berrylike drupes grow in compact pyramidal clusters. Each fruit is globular and crowned with remnants of calyx and style. The inclosed seedlike nutlets number from three to five. June.

Leaves. — The opposite leaves are compound with five to seven ovate-lanceolate leaflets. These are finely toothed and acute.

Flowers. — The small cream-white flowers, with their pale yellow stamens, grow in compound pyramidal cymes. April, May.

For two weeks I had been looking, without success, for the bright red berries which this Elder bears. When almost in despair over securing a specimen, I chanced to be trolleying in the vicinity of Mt. Tom, when my eye suddenly caught a gleam of red against a rocky background. I knew at once that it was my coveted prize. Fortunately a switch was near, and while the car waited there I was able to hurry back, get my specimen, and resume my journey. This especial plant was growing out of a wall of rock. In general, it is found in rocky woodlands and has a wide range from New Brunswick south to Georgia and westward across the continent. A variety with white berries is said to have been found in the Catskill Mountains.

The shrub grows from two to twelve feet high. The older stems are brown and warty. In blossom and in fruit the plant may be readily distinguished from the Common Elder, and at other times the brown pith in the young shoots serves as a determining feature. The fruited shrub, at a distance, looks something like a sumac.

HOBBLE BUSH. WAYFARING TREE

Viburnum alnifolium. Viburnum lantanoides
Honeysuckle Family

Fruit. — The large ovate drupes are coral-red, turning later almost black. Each contains an oblong-oval nut, which is obtusely pointed and grooved on both sides. The drupes grow in scanty clusters.

Leaves. — The leaves are large, light green, heart-shaped, abruptly pointed, sharply toothed, and have rusty wool on the veins beneath. In the fall the leaves turn to red and orange shades.

Flowers. — The flowers are in broad, showy cymes with larger, showy, usually sterile flowers around the margin. May.

The reclining branches of this shrub often take root, making loops which "trip the wayfarer." "Hobble Bush" is a name which is suggested by the appearance of the plant with its looping branches. It grows in low, moist woods from New Brunswick to Ontario and south to Pennsylvania and in the mountains to North Carolina.

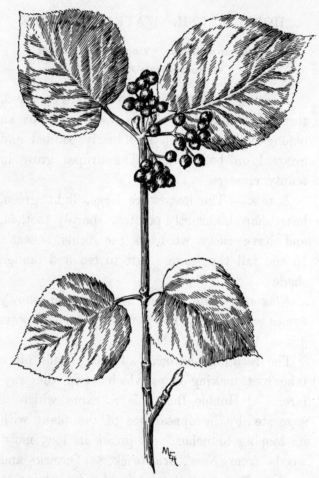

Hobble Bush (*Viburnum alnifolium*)

CRANBERRY TREE (*Viburnum Opulus*)

CRANBERRY TREE.   GUELDER ROSE

Viburnum Opulus   **Honeysuckle Family**

Fruit. — The beautiful globose drupes grow in terminal cymes. They are bright red when ripe, having changed from green to greenish yellow and yellowish red. The separate fruits are about the size of Choke Cherries. At the tip are traces of the calyx teeth. The fruits are fleshy and inclose a flat stone with a thin crustaceous coat. The stone is without furrows or grooves. The fruit is acid and a trifle bitter. August, and persistent.

Leaves. — The leaves are opposite. They are strongly three-nerved and three-lobed. They are sparingly toothed, being usually entire along the margins of the sinuses. The developed leaves are dark green above and paler beneath. Dull red or purple are the autumn colors.

Flowers. — The white flowers grow in a flat cyme. The marginal ones are larger and neutral, the central ones smaller and perfect. June, July.

This is an interesting and attractive shrub. In the spring, it bears its showy white flower clusters, with their margins of large flowers, "just for show," to attract the roving insect to the encircled blooms. By cultivation of the European form, the central flowers have been changed to large neutral ones like those at the margin, the flower head has become spherical, and the Snowball Tree of the garden is the result. The wild shrub not only yields a graceful bloom, but is most attractive in fruit, with its erect clusters of brilliantly colored drupes. These lose somewhat of their brilliancy after frost, but are conspicuous throughout the winter. The fruits make a good jelly and an agreeable sauce.

The plant extends north from Pennsylvania to New Brunswick and west to Michigan, South Dakota, and Oregon.

FEW–FLOWERED CRANBERRY TREE

Viburnum pauciflorum   **Honeysuckle Family**

This species is fewer flowered than the preceding and lacks the marginal neutral flowers. The fruit clusters are small. The drupes are light red and contain flat, scarcely grooved stones.

It grows in cold mountainous woods nearly throughout Canada, in New England, and Pennsylvania, and in the Rockies in Colorado and Washington.

TINKER'S WEED. WILD OR WOOD IPECAC. WILD COFFEE. HORSE GINSENG. FEVERWORT

Triosteum perfoliatum   **Honeysuckle Family**

Fruit. — The rather dry orange or scarlet drupes are borne at the junction of leaf and plant stems. The long lobed calyx remains attached to the fruit summit. The drupes are covered with fine hairs and inclose three bony nutlets.

Leaves. — The leaves are ovate to broadly oval, acute at the apex, abruptly or gradually narrowed at the base, and stemless or united about the stem. They are soft pubescent beneath and somewhat hairy above.

Flowers. — The purplish brown flowers are usually clustered. June.

This is a coarse hairy herb, growing from Canada and New England southward to Iowa and Alabama.

INDIAN CURRANT. CORAL BERRY

Symphoricarpos Symphoricarpos. Symphoricarpos vulgaris
Honeysuckle Family

Fruit. — The fruit varies in ripening from coral-red to reddish purple. It is small, ovoid-globose, and bears the calyx teeth at the summit. The skin is thin; the flesh is dry, mealy, and insipid; and, although there are four cells, the seeds are but two in number, two ovules being abortive. The seeds are white and hard. The berries grow in clusters from the axils of most of the leaves. The fruits persist during the winter.

Leaves. — The entire oval or ovate leaves are on short stems and opposite. They are a dull green and somewhat hairy beneath. They are usually obtuse both at apex and base.

Flowers. — The pinkish bell-shaped flowers are somewhat hairy at the throat. They grow in clusters in the leaf axils. July.

This plant is most prolific in fruit, which persists after the leaves have dried and fallen. The clusters extend nearly the length of the stem and bend it with their weight.

INDIAN CURRANT (*Symphoricarpos Symphoricarpos*)

143

The plant grows wild in New Jersey and Pennsylvania, along the banks of the Delaware River, in New York, and west to Dakota. It reaches Georgia and Texas on the south. It is often cultivated and sometimes escapes.

AMERICAN WOODBINE. ITALIAN OR PERFOLIATE HONEYSUCKLE

Lonicera Caprifolium.   Lonicera grata
Honeysuckle Family

Fruit. — The red berries are in a sessile terminal cluster, subtended by a united pair of leaves. The calyx teeth crown the summit of the several-seeded fruit.

Leaves. — The two or three upper pairs are united by their bases. The lower ones are without stems or have very short stems. They are obovate or oval and entire.

Flowers. — The fragrant flowers are whitish with purple tubes. They are strongly two-lipped.

New Jersey and Pennsylvania are the northern limits of this climbing vine in its wild state. It is often cultivated, and sometimes escapes.

HAIRY HONEYSUCKLE

Lonicera hirsuta   **Honeysuckle Family**

Fruit. — The red berries grow in short terminal spikes. The calyx teeth are at the summit and the berry is several-seeded.

Leaves. — The leaves are large and have hairy margins and under leaf surfaces. The base is rounded or narrowed and the apex obtuse. One or two upper pairs have united bases, the others are stemless or have very short stems.

Flowers. — The orange-yellow blossoms grow in short interrupted spikes. They are two-lobed and the tube is clammy-pubescent. July.

This twining shrub is a coarse species, with large leaves and hairy branches, leaves, and flowers. It grows from Maine to Pennsylvania, and west to Michigan and Minnesota.

SMOOTH–LEAVED OR GLAUCOUS HONEY-SUCKLE

Lonicera dioica.   Lonicera glauca
Honeysuckle Family

Fruit. — The berries form a compact cluster, composed of a series of usually three whorls.

The whorls are more or less imperfect, owing to the nondevelopment of some of the berries. The cluster is borne on a short terminal stem. Each fruit is several-seeded and has persistent calyx teeth at the summit. The berries are without stems. They vary in color from orange to red, the red ones being the ripest. They are covered with a bloom. The pulp is similar in color to the skin. The berry is translucent. July, August.

Leaves. — The leaves are mostly oblong and from two to three inches in length. The bases of the one to four upper pairs are united. The leaves not united by their bases are stemless. The terminal pair varies in shape from oblong to oval, and with its rich green coloring forms a most attractive setting for the bright berry cluster which it surrounds. The leaf margins are entire and the under surface is whitened.

Flowers. — The flowers grow in terminal clusters. They are greenish yellow, sometimes tinged with red. The tube expands into two lips, the lower one narrowed and the upper one broader and four-lobed. The inside of the tube, the style, and the bases of the filaments are hairy.

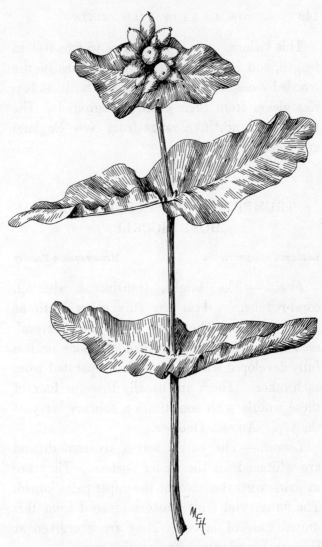

SMOOTH-LEAVED HONEYSUCKLE (*Lonicera dioica*)

This twining vine is from three to five feet in length, and is a most attractive feature of the wooded roadside in July, when the brilliant berries gleam from their green background. The plant has a northern range from New England and Pennsylvania.

TRUMPET HONEYSUCKLE.   CORAL HONEYSUCKLE

Lonicera sempervirens   **Honeysuckle Family**

Fruit. — The bright, translucent, shining, coral-red berries bear the tiny calyx teeth at the summits. They are ovoid and several-seeded and grow in a spike of more or less fully developed whorls, somewhat separated from each other. There are usually three or four of these whorls with sometimes a solitary berry at the top. August–October.

Leaves. — The entire leaves are smooth and are whitened on the under surface. They are in pairs, with the bases of the upper pairs joined. The flower and fruit clusters proceed from this united pair of leaves. They are evergreen at the south and deciduous at the north.

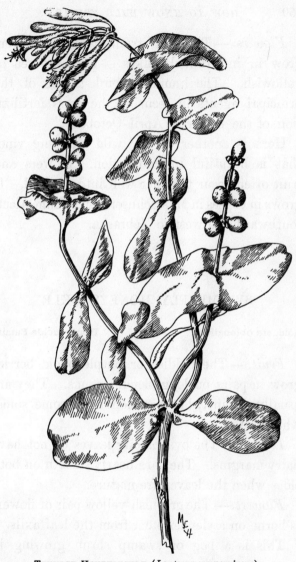

TRUMPET HONEYSUCKLE (*Lonicera sempervirens*)

Flowers. — The long trumpet-shaped flowers grow in interrupted spikes. They are red or yellowish. The humming bird is one of the principal agents in securing the cross fertilization of the flowers. April–October.

Here is another of our wild climbing vines that is beautiful in cultivation. Flowers and fruit often occur together well into the fall. It grows in copses in Massachusetts and Connecticut, southward and west to Nebraska.

SWAMP FLY HONEYSUCKLE

Lonicera oblongifolia   **Honeysuckle Family**

Fruit. — The reddish or purple ovoid berries grow in pairs on long slender stems. They are usually distinct, but occasionally become somewhat united.

Leaves. — The oval-oblong leaves do not have hairy margins. They are nearly smooth on both sides when the leaves are mature.

Flowers. — The greenish yellow pair of flowers is borne on a slender stem from the leaf axils.

This is a bog or swamp shrub growing in

northern New England and New York, and
west to Minnesota.

AMERICAN FLY HONEYSUCKLE

Lonicera ciliata   **Honeysuckle Family**

Fruit. — Two globular or ovoid red berries
are borne on the same stem, which grows from
the leaf angles. The berries are not united at
the base, and each bears at the summit minute
calyx teeth. The berries are several-seeded.
The bracts at their bases are minute. June.

Leaves. — The thin, light green leaves are
oblong-ovate, somewhat rounded or heart-shaped
at the base and acutish at the apex. They are
opposite and have hairy margins.

Flowers. — The yellowish green, five-lobed
flowers grow in pairs on a slender stem.

This is a straggling shrub from three to five
feet high. The stems are brownish. It grows
in rocky woods from New Brunswick to Pennsylvania and west to Minnesota.

BLACK OR DARK PURPLE

BLACK OR DARK PURPLE

WHITE CLINTONIA

Clintonia umbellulata   **Lily-of-the-Valley Family**

Fruit. — The berry of the White Clintonia is black, not quite so large as that of the Yellow, and has few seeds. The fruits grow in umbels at the top of the hairy stem, which sometimes bears also a single small leaf.

Leaves. — The leaves are oval or oblong, with hairy margins. The stalks of the two to four leaves sheathe the base of the flower stem.

Flowers. — The flowers are white, often speckled with green or purple. They are fragrant.

This species is confined to rich woods in the Alleghany Mountains from New York to Georgia.

STAR–FLOWERED SOLOMON'S SEAL

Vagnera stellata.   Smilacina stellata
Lily-of-the-Valley Family

Fruit. — The few berries grow in a terminal cluster. They are, according to Mathews, spotted at first and later becoming dull red. Gray says they are blackish, and Britton and Brown that they are green with six black stripes, or black. They are rather larger than the fruits of Wild Spikenard, quite hard and opaque.

Leaves. — The oblong-lanceolate leaves are slightly clasping. The apex is acute or blunt. The leaf is flat or a trifle concave.

Flowers. — The white starlike flowers grow in a small terminal raceme.

This is a smaller species than *V. racemosa*, seldom growing over a foot high. Its rootstock is rather slender. It favors banks of streams and moist meadows. It extends south to New Jersey and west to Kansas, and, according to Britton and Brown, to California.

HAIRY SOLOMON'S SEAL

Polygonatum biflorum **Lily-of-the-Valley Family**

Fruit. — The berry is nearly black, with a bloom. It is pulpy, three-celled, with one or two seeds in each cell. The stigma is at the summit. The berries grow on slender, drooping stems from the axils, and are solitary, or two in a cluster, rarely three. August, September.

Leaves. — The light green leaves are oblong-ovate, alternate, and sessile. They are parallel-ridged and acute at the apex. The under surface is whiter and hairy.

Flowers. — The pale green flowers look like tassels hanging in drooping clusters of from one to three flowers from the leaf axils. May.

The scars left on the thick horizontal root-stocks, where the stalks of preceding years grew, give rise to this plant's common name, Solomon's Seal. These marks, which are indicative of the age of the root, are somewhat like the impression of a wax seal. This is a graceful, low, wood plant, with a curving stem and drooping flower and fruit clusters.

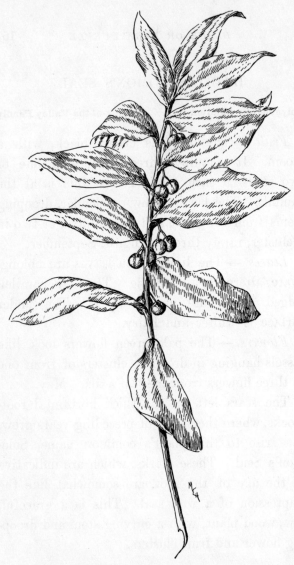

HAIRY SOLOMON'S SEAL (*Polygonatum biflorum*)
158

SMOOTH SOLOMON'S SEAL

Polygonatum commutatum.   Polygonatum giganteum
Lily-of-the-Valley Family

Fruit. — This globular berry is also nearly black with a bloom.   It is larger than the preceding, in keeping with the larger proportions of the species.   The clusters vary in the number of their fruits from one to six.   These grow on long, stout, drooping stems from the leaf axils. The berry is three-celled, one cell sometimes containing six seeds.   August, September.

Leaves. — The large leaves are ovate and partly clasping.   They are smooth throughout, rather darker green than the smaller species and somewhat paler beneath.   The yellow fall leaves contrast well with the dark berries.

Flowers. — The drooping jointed peduncles bear two to eight large, greenish, bell-shaped flowers.   June.

The tall, stout stalks, sometimes seven feet high, with their large, spreading, gracefully curved leaves and the numerous nodding clusters of black balls are imposing additions to the flora of moist roadsides.   They also abound along

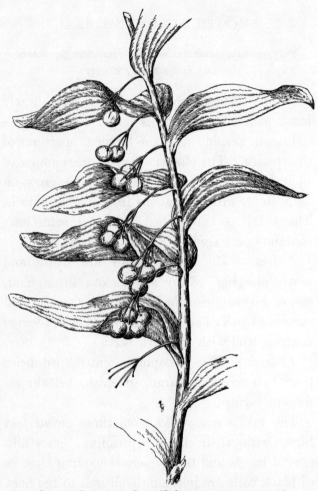

SMOOTH SOLOMON'S SEAL (*Polygonatum commutatum*)

streams. The species grows as far west as the Rocky Mountains.

INDIAN CUCUMBER ROOT

Medeola Virginiana   **Lily-of-the-Valley Family**

Fruit. — The dark purple berries are borne at the summit of the plant on upright stems. They are globular, usually three or four in number, three-celled, and few-seeded. The mark of the style is at the tip. September.

Leaves. — The leaves are in two whorls. The lower whorl is borne about midway of the stem and consists of from five to nine obovate-lanceolate leaflets, which are stemless, parallel-ribbed, and netted-veined. The upper whorl, at the top of the stem, is usually of three, occasionally more, smaller ovate leaflets.

Flowers. — The greenish yellow flowers are borne on drooping stems and are often nearly hidden beneath the upper whorl of leaflets. They are like small lilies and have recurved perianths, six recurved reddish stamens, and three recurved stigmas. June.

The flowers on their drooping stems are often tucked under the upper leaflets, which serve as

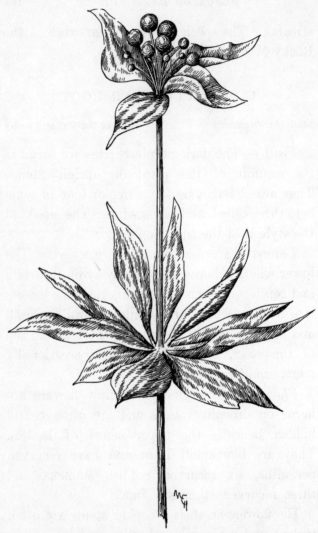

INDIAN CUCUMBER ROOT (*Medeola Virginiana*)

umbrellas for them. When the plant is fruiting the stems become erect, and, in the fall, when the berries are ripe, the crowning leaflets and fruit stems, tinged with dull reds, serve as signals to the birds that harvest time has come.

The horizontal tuberous rootstock is a characteristic feature. It is white, and similar in taste to a cucumber.

"Its white tuberous root is crisp and tender, and leaves in the mouth distinctly the taste of cucumber. Whether or not the Indians used it as a relish I do not know." — BURROUGHS.

CARRION FLOWER

Smilax herbacea   **Smilax Family**

Fruit. — The flattened globose berries grow in more or less full rounded clusters. They are borne on long peduncles, sometimes six or even eight inches long. These stems grow from the axils of single leaves or from the axils of leafy branches, which themselves spring from the leaf axils. The berries are black with a bloom which is often decidedly blue. The flesh is thin, and the berry variable in the number of its seeds,

two to six. It is normally three-celled, with
two seeds in each cell. August, September.

Leaves.—The simple alternate leaves are seven
to nine ribbed and are netted veined. They are
usually round-ovate. The apex is acute, sometimes bristle-pointed, and the base is heart-
shaped or obtuse. The leaves are entire. A
pair of tendrils proceed from the leaf stem.
The under surface is lighter than the upper.

Flowers.—The diœcious greenish flowers grow
in from twenty- to forty-flowered clusters. They
are ill-scented, like "a dead rat in the wall," as
Thoreau describes it. They are fertilized by
insects, especially carrion-loving ones.

The main stem is neither woody nor thorny.
By means of the numerous tendrils it climbs
over any and every support it may encounter.
The spherical clusters of bluish black fruits are
very attractive about the middle of August.
They are a frequent sight amidst the roadside
flower tangles, and flourish along streams and
in moist places. They range east from Minnesota, Missouri, and Texas to the Atlantic.

CARRION FLOWER (*Smilax herbacea*)

165

HALBERD–LEAVED SMILAX

Smilax tamnifolia   **Smilax Family**

Fruit. — The fruit is a berry similar to that
of the Carrion Flower, but smaller. The clusters
are rather small and the peduncles shorter than
in preceding species. The berry is from one- to
three-seeded.

Leaves. — The leaves are broad at the base and
narrow decidedly about the middle of the leaf,
making the base almost lobed. They are thick,
leathery, and green on both sides.

This, like several other smilax species, has its
northern range in dry or sandy portions of New
Jersey. It extends south to South Carolina and
Tennessee. It is unarmed, and usually has circular stems and branches.

GLAUCOUS–LEAVED GREENBRIER
FALSE SARSAPARILLA

Smilax glauca   **Smilax Family**

Fruit. — The globose black berries grow in umbels on flattened peduncles, which are rarely twice
the length of the petioles. The umbels spring

from the leaf axils. The flesh is thin and incloses from one to three large seeds. September.

Leaves. — The oval leaves vary in width. They are conspicuously whitened beneath. The apex is rather obtuse but ends in a sharp point. The petioles are rather short and bear tendrils near their bases. The leaves are somewhat persistent.

Flowers. — The small greenish yellow flowers are diœcious. There are from six to twelve blossoms in the flower umbel, which is borne on a flattened stem.

This woody vine sometimes bears scattered prickles, sometimes none. The stem is circular. It grows in thickets from Massachusetts to Florida and extends west to Texas, Missouri, and Indiana.

GREENBRIER. CATBRIER. HORSEBRIER

Smilax rotundifolia   **Smilax Family**

Fruit. — The globular blue-black berries are covered with a bloom. They grow in umbels on a flattened stem, which seldom exceeds in length that of the petiole. They are one- to three-seeded. September and often persistent.

Leaves. — The leaves are ovate or round-ovate, somewhat heart-shaped or rounded at the base, and abruptly pointed at the apex. They are leathery, shining when mature, green on both sides, entire, and smooth. Tendrils grow from the leaf stems and are modified forms of stipules.

Flowers. — Small, yellowish green, diœcious flowers grow in rather small clusters on short cluster-stems. April–June.

This Greenbrier is quite common. Its yellowish green stem is round and the branches are somewhat four-angled. It sometimes grows as long as forty feet, and is generally armed throughout with stout prickles. It grows in moist places from New England to Georgia and as far west as Minnesota.

HISPID GREENBRIER

Smilax hispida   **Smilax Family**

Fruit. — The bluish black berries are in umbels, borne on stems that are over twice as long as the leaf stems. They are one- to three-seeded.

Leaves. — The egg-shaped leaves are thin, green on both sides, rounded or slightly heart-shaped at the base, and pointed at the apex.

Flowers. — The umbels are composed of flowers somewhat larger than those of the Catbrier.

This vine is distinguished by the long black bristles which densely cover the lower portion of the stem. The upper portion is generally unarmed. Connecticut is its northern boundary. It grows in moist thickets.

LONG–STALKED GREENBRIER

Smilax Pseudo-China   **Smilax Family**

Fruit. — The umbels of black berries grow on flattened stems from the leaf axils. They are quite full, bearing eight to sixteen berries in a cluster. The peduncles are considerably longer than the petioles, being from one to three inches long.

Leaves. — The firm, almost leathery leaves are green on both sides. They are ovate or sometimes nearly lobed at the base. The apex is acute or bristle-pointed, and the edge is sometimes roughened with fine bristle-like teeth.

Flowers. — The flowers are diœcious and grow in full-flowered clusters on long flattened stems. July.

Long-stalked Greenbrier, as the name indicates, bears its flowers and fruits on long stems. These are more than twice as long as the petioles. The main stem is circular and sometimes armed with straight prickles along its lower part. Most of the plant is unarmed. It belongs to our southern section from New Jersey to Florida and west to Indiana and Missouri. It favors dry sandy soil.

BRISTLY GREENBRIER. STRETCH BERRY

Smilax Bona-nox   **Smilax Family**

Fruit. — These berries are black with a bloom. The umbels are borne on a flattened stem about twice the length of the leaf stem. The berry usually has but one large seed. Its pulp is elastic, hence the name Stretch Berry.

Leaves. — The leaves are round, heart-shaped, or often much broadened at the base and narrowed midway of their length, giving a somewhat two-lobed appearance to the base. The apex is bristle tipped, and the margin and mid·

rib are often spiny. The upper and lower surfaces are both green and shining. The stems are tendril-bearing. The leaves cling long to the vines.

Flowers. — The diœcious flowers grow in quite full umbels. April–July.

The prickles on this vine are few, short, stiff, and scattered. The circular stem and the angular branchlets are both green. It has been reported in Massachusetts, and extends from New Jersey to Florida, and west to Illinois, Texas, and Missouri.

HACKBERRY. SUGARBERRY

Celtis occidentalis   **Elm Family**

Fruit. — The solitary drupe is about as large as a pea. It grows from the leaf axil on a drooping stem. The calyx is persistent, and the stigma is at the tip. The ripe fruit is dark purple. The flesh is rather thin and very sweet, and the stone is large. September, October, and persistent.

Leaves. — The two sides of the leaf are quite unlike, one being much broader at the base than the other, which looks as if a piece of it had been cut off obliquely. The apex is pointed. The margin is toothed except at the base. Autumnal coloring is yellow.

This tree is similar to an elm in appearance. Its fruit is much appreciated by the winter birds. The trees which I have seen have been much disfigured by numerous insect galls upon the leaves. Its range is in woods and along river banks, in New England southward and west to Minnesota.

RED MULBERRY

Morus rubra   **Mulberry Family**

Fruit. — The fruit seems at first glance to resemble a Blackberry in structure. It differs, however, in being the product of a spike of several flowers instead of the development of several carpels of the same flower. Each separate fruit consists of an achene or nut surrounded by the calyx lobes which have become juicy. Each achene bears at the summit the tips of the two styles. Only one of the two ovaries of the flower develops. The multiple or collective fruit formed by the crowding together

of the separate fruits is about an inch long, is
sweet, juicy, and edible. It grows on a short
stem, usually from the axil of the leaf. The
fruit in ripening changes from green to red to
dark purple. July.

Leaves. — The leaves are variable in shape but
are usually heart-ovate. On young shoots they
are often lobed. The margins are coarsely
toothed. The upper surface is shining and may
be smooth or rough. The lower surface is
lighter. Yellow is the autumnal color.

Flowers. — A tree sometimes bears both staminate and pistillate clusters of flowers, and sometimes but one kind. A few pistillate flowers
are occasionally found in the staminate flower
spikes.

The Red Mulberry is the only species native
to America. The tree does not usually attain a
great size, but sometimes reaches a height of
from sixty to seventy feet. The finest trees are
to be found along the lower Ohio and the Mississippi rivers. They range from Massachusetts to
Florida and west to Kansas and Nebraska. The
leaves do not serve successfully as food for silkworms. These flourish best on the White Mulberry leaves. An interesting feature occurs in

RED MULBERRY (*Morus rubra*)

175

connection with the pollination of the flowers. At the precise time that the anthers are ready to open, the filaments uncoil like a spring and throw the pollen upon the breezes.

POKE. SCOKE. GARGET. PIGEON BERRY

Phytolacca decendra   **Pokeweed Family**

Fruit. — The dark purple berries grow in long lateral racemes opposite the leaves. The berries are like a sphere flattened vertically and are from five- to twelve-celled. Each cell contains one vertical seed. The berry is filled with a crimson juice. The calyx persists at the base. September.

Leaves. — The large coarse leaves are often veined with purple.

Flowers. — The five sepals are white or pinkish and surround the conspicuous green ovary. The corolla is lacking.

This is a large rank perennial. The large roots are poisonous but the young plants are cooked in early summer for " Greens," and are considered almost equal to Asparagus. The sturdy plants often occur along the roadside, and I have seen a rocky hillside pasture overgrown

with them.   Birds of several different kinds eat
the berries.

POKE (*Phytolacca decendra*)

"Pokeweed is a native American, and what a
lusty, royal plant it is!" — BURROUGHS.

CANADA MOONSEED

Menispermum Canadense   **Moonseed Family**

Fruit. — The ovary is nearly straight, and has the stigma at the apex. In the development of the fruit an incurving takes place, bringing the stigma mark near the base of the fruit. This gives the stone the form of a crescent or ring; hence the name Moonseed, because of its crescent shape. The stone is flattened laterally, and is wrinkled and grooved. The drupes are globose-oblong, one-seeded, black with a bloom. They grow in loose clusters and resemble Frost Grapes in appearance. The clusters, however, grow from the leaf axils instead of opposite them. September.

Leaves. — The leaves are broad-ovate with usually three to seven lobes. They are heart-shaped at the base, and have a pale under surface. The leaf stem is slender and usually attached within the edge of the leaf.

Flowers. — The small, greenish white, diœcious flowers grow in loose clusters from the leaf axils. June, July.

This woody climber, sometimes twelve feet

long, is readily distinguished from a grapevine, which it somewhat resembles, by the axillary position of flower and fruit cluster. It is common along streams south to Georgia and Arkansas.

WILD GOOSEBERRY

Ribes Cynosbati   **Gooseberry Family**

Fruit. — The brownish red berry usually grows singly. It is prickly, occasionally smooth, and has numerous seeds with crustaceous coats inclosed in gelatinous ones. The seeds are suspended by tiny threads in a pulpy mass. The berry bears at the summit the shriveled remains of the calyx. The flavor is good, but the sharp, awl-shaped prickles are objectionable. August.

Leaves. — The three- to five-lobed leaves are alternate or clustered. The base is heart-shaped, and the lobes are incised or serrate. One or more spines are usually found at the base of the petioles.

Flowers. — The bell-shaped greenish flowers grow singly or in a few-flowered raceme.

This is a low shrub of rocky woods from New Brunswick south, especially along the mountains, to North Carolina. It extends west to Manitoba and Missouri.

WILD BLACK CURRANT

Ribes floridum   **Gooseberry Family**

Fruit. — The drooping racemes bear smooth, black, round-ovoid berries with a linear bract at the base of each pedicel. The dried calyx is at the top of each fruit. Each berry is many-seeded, with the seeds attached by tiny threads to two opposite lateral placentæ. The fruit is watery and insipid. July, August.

Leaves. — Tiny resinous dots on the leaves are characteristic. The five to seven lobes of each leaf are doubly toothed.

Flowers. — The large whitish flowers grow in drooping, loosely-flowered, bracted, and downy racemes.

This shrub is erect, and reaches a height of from three to five feet. It is found in woods from Nova Scotia south to Virginia and west to Kentucky, Iowa, and Nebraska.

BLACK RASPBERRY (*Rubus occidentalis*)

WILD RED RASPBERRY (*Rubus strigosus*)

BLACK RASPBERRY. THIMBLE BERRY

Rubus occidentalis   **Rose Family**

Fruit. — The small, black, juicy drupes are
packed in diminishing circles about the elongated
receptacle, forming a flattened hemispherical,
aggregate fruit. This separates when ripe from
the receptacle and the reflexed calyx lobes at the
base. Each drupelet is woolly near its points of
contact with other drupelets. The remainder of
its surface is smooth and shining. The fruits
grow in small terminal clusters. The peduncles
are set with recurved prickles. The fruit is
sweet and delicious in flavor. July.

Leaves. — There are usually three leaflets to
each compound leaf. The two lateral ones have
short stems. Scattered prickles are on the leaf
stems. The leaflets are much whitened beneath,
coarsely doubly toothed, and acutely pointed.

Flowers. — The white blossoms grow in terminal corymbs.

Pull off one of the fruits from the receptacle,
slip it over the tip of the little finger, and see if
it is not a veritable " finger cap," worthy of its
name Thimble Berry. The plant yields, each

year, beside the fruiting stems, gracefully curved
leafy shoots, which will be the fruit-bearing portion of next year's growth.   The Raspberry has
been generally cultivated.

In its native haunts it is found drooping over
rocks, growing in clumps about decaying stumps
or trees and in fence rows.

BLACKBERRIES

The Blackberry group seems to be especially
difficult of exact classification.   L. H. Bailey, in
his book on " The Evolution of our Native
Fruits," names certain varieties which are recog-
nizable even by a non-expert, and I have de-
parted somewhat from Britton and Brown's
classification to adopt Bailey's more recent
nomenclature and divisions.

He separates, first, the Blackberries from the
Trailing Blackberries, or Dewberries.

The Dewberries are distinguished by their
trailing habit of growth, their custom of rooting
from the tips, and by the few scattered flowers
in the flower cluster, the central one of which
blossoms first.

The Blackberry fruit, in general, is a collection of small drupes, which remains attached when ripe to the juicy white receptacle on which it is borne.

Our principal Dewberries are two in number : Low Running Blackberry and Running Swamp Blackberry.

LOW RUNNING BLACKBERRY

Rubus villosus. Rubus Canadensis   **Rose Family**

Fruit. — The fruit grows in small clusters. It is usually hemispherical or ovoid. The drupelets are large, juicy, and rather sour until fully ripe, when they are quite sweet. At the base of each fruit is the calyx, from which the berry separates when it falls, leaving many dried stamens visible in the calyx cup.

Leaves. — The leaves have three to seven oval or ovate leaflets, which are sharply doubly toothed. They are quite thick and large. Leaf-like bracts grow on the flower and the fruit clusters.

This is the common Dewberry of the north, and is a frequent roadside trailing vine. The species is very variable.

The rich dark reds of its fall leaves, spreading

over stone walls, rocks, roadsides, and pasture lands, and contrasting with the yellow of the Golden-rod and the blue of the Aster, are important factors in the autumnal color scheme.

The *Rubus villosus*, that was named and described in 1789, has long been taken by bot-

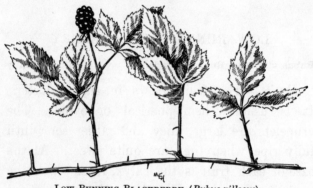

LOW RUNNING BLACKBERRY (*Rubus villosus*)

anists to be the High-bush Blackberry. In 1898, Bailey, after personally examining the specimens described by Aiton as *Rubus villosus*, decided that they were specimens, not of the High Blackberry but of the northern Low Blackberry or Dewberry. To this plant, then, the name *Rubus villosus* rightfully belongs.

> " For my taste the blackberry-cone,
>   Purpled over hill and stone."
>   — WHITTIER'S *Barefoot Boy.*

RUNNING SWAMP BLACKBERRY

Rubus hispidus   **Rose Family**

Fruit. — The fully ripened fruit is nearly black. It is small, consisting of but few grains.

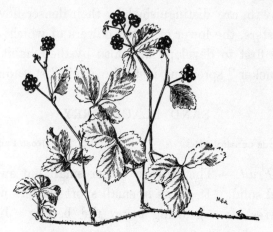

RUNNING SWAMP BLACKBERRY (*Rubus hispidus*)

The berries are borne on leafless stems which are often bristly. August.

Leaves. — The small, usually three, obovate leaflets are smooth, coarsely toothed, and blunt at the tip. They are shining and firm and appear evergreen.

Flowers. — The flowers are small and white, with few in the cluster.

The fall foliage is brilliant. It grows not only in swamps, but in sandy places as well.

The Blackberries, aside from their more erect growth, are distinguished by their denser flower clusters, the lower or outer flowers of which are the first to develop; and also by their habit of " sucker " spreading instead of " tip " spreading.

SAND BLACKBERRY

Rubus cuneifolius   **Rose Family**

Fruit. — The fruit is rather small but sweet and solid. It grows in small short clusters with leaves often growing below the berries. July, August.

Leaves. — The compound leaves consist of from three to five leaflets, which are obovate, obtuse, toothed, and quite thick. They are dull green above, whitened and woolly beneath.

Flowers. — The white or pinkish flowers grow in short, usually terminal clusters of few flowers.

This is a low variety, from one to three feet

high. It is stiff and armed with stout prickles. It favors sandy soil.

Of the High-bush Blackberries, Bailey makes three divisions: *Rubus nigrobaccus*, *Rubus argutus*, and *Rubus Canadensis*. Professor Porter also describes another form which Bailey is inclined to accept as a separate one, *Rubus Allegheniensis*, or Mountain Blackberry.

COMMON OR HIGH–BUSH BLACKBERRY

Rubus nigrobaccus.   **Rubus villosus**   **Rose Family**

Fruit. — These so-called berries are oblong, seedy, firm, and sweet. They grow in long loose clusters, the lower berries usually ripening first. The five long, narrow calyx lobes are reflexed at the base. July, August.

Leaves. — The leaflets are three or five in number. Each has a distinct stem, the terminal one having the longest stalk. The leaflets are ovate, pointed, and coarsely serrate. The under-leaf surface is hairy and glandular.

Flowers. — The large white flowers are borne in long clusters. Each pedicel is long and forms

High-bush Blackberry (*Rubus nigrobaccus*)

a broad angle with the axial stem. The flower stems are hairy and glandular.

The stems of the High Blackberry are furrowed, often recurved, and bear stout, hooked prickles. They are sometimes ten feet high. This is the common High Blackberry, and is found in woods and along country roads and fence rows. Iowa, Missouri, and Kansas are its western limits and the North Carolinian Mountains its southern.

Through a confusion in the identification of early specimens, the name *Rubus villosus* has been applied to the High-bush Blackberry instead of to the Low Blackberry, which Gray calls *Rubus Canadensis*. This latter term, however, belongs to the Thornless Blackberry. The Low Blackberry must bear its rightful title, *Rubus villosus*, and this leaves the High Blackberry without a name. Bailey has christened it *Rubus nigrobaccus*.

The variety *sativus*, Short Cluster Blackberry, has rounder fruits, that grow in short clusters. The drupelets are loose and large.

The leaflets are broader and the apex blunter.

This is the common High Blackberry of the open fields. It is not so tall as the type.

MOUNTAIN BLACKBERRY

Rubus Allegheniensis   **Rose Family**

Fruit. — The drupelets are small and dry, and form a long, narrow fruit, with tapering top. The flavor is spicy.

Leaves. — The smaller teeth and long drawn-out apex are distinctive features.

This species has reddish branches and leaf stems. It is considered by some to be a mountain form of *Rubus nigrobaccus*.

LEAFY CLUSTER BLACKBERRY

Rubus argutus   **Rose Family**

Fruit. — This species is distinguished by a shorter, leafy fruit cluster.

Leaves. — The leaflets are smaller and narrower, somewhat rigid, nearly smooth, and coarsely serrate.

It is a lower species than *R. nigrobaccus*, stiff, straight, and nearly smooth or quite so. It is distinctly a southern species, taking the place there which is occupied in the north by *R. nigrobaccus*. It has a wide range.

THORNLESS BLACKBERRY

Rubus Canadensis   **Rose Family**

Fruit. — The fruit ripens late and is sweeter than the other blackberries.

This species is distinguished by smooth leaves and stems and its usual lack of thorns. The leaves are long, narrow, and have a long acumination. The leaf stems are long and slender and the stipules are long.

It grows as far south as North Carolina.

BLACK CHOKEBERRY

Aronia nigra. Pyrus arbutifolia, Var. melanocarpa

Apple Family

This plant is smoother than *Aronia arbutifolia*, but the chief difference is in the fruit, this pome being larger, more juicy, and black in color.

The shrub often grows in the vicinity of Huckleberry Bushes, and the two fruits somewhat resemble each other. I well remember the cautions which as a child I received against mixing the two, the Chokeberry being con-

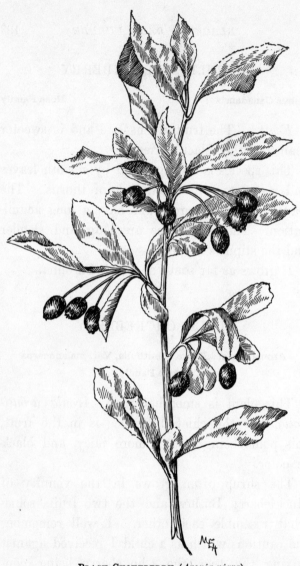

BLACK CHOKEBERRY (*Aronia nigra*)

194

sidered poisonous. I distinguished between
them by the red juice or flesh of the "Dog-
berry," as we called it. So strong was my
early belief in its "killing" qualities, that,
despite the testimony of several books as to its
sweetness, pleasant flavor, etc., I confess I test
them rather gingerly. The range is nearly the
same as that of the Red Chokeberry.

OBLONG-FRUITED JUNEBERRY

Amelanchier oligocarpa   **Apple Family**

Fruit. — The pear-shaped pome is dark purple,
and covered with a thick bloom.

Leaves. — The thin oblong leaves are narrowed
at each end and often acute. They are sharply
toothed.

Flowers. — There are few, one to four, blossoms in a raceme. The pedicels are rather long
and slender.

This is a low, nearly smooth shrub, growing
in wet places in Ontario, New England, and
along the shores of Lake Superior.

PORTER'S PLUM

Prunus Allegheniensis   **Plum Family**

Fruit. — The dark purple globose drupe is small, only about half an inch in diameter. It has a bloom. The flavor is pleasantly acid. The stone has a groove on one side and a slight elevation on the other. August.

Leaves. — The lanceolate or ovate-oblong leaves are finely toothed, and often have a long pointed apex.

The flowers resemble those of *Prunus Americana.*

This is a low tree or straggling shrub of the Alleghany Bluffs. It seldom bears thorns.

SLOE. BLACKTHORN

Prunus spinosa   **Plum Family**

Fruit. — The globose drupe is about half an inch in diameter. It is black, with a bloom, and grows singly or in pairs from the sides of the branches. The stone has one sharp edge.

Leaves. — The ovate or oblong leaves are obtuse at the apex and narrowed at the base.

They are sharply toothed and smooth when mature.

Flowers. — The white flowers are solitary or two in a cluster. April, May.

This thorny shrub was introduced from Europe, and often occurs as an escape along roadsides from Massachusetts to New Jersey and Pennsylvania.

A variety, *insititia*, Bullace Plum, is less thorny, and has hairy leaves and stems.

SAND CHERRY. DWARF CHERRY

Prunus pumila   **Plum Family**

Fruit. — The drupes are about half an inch long and are dark red or, when fully ripe, black. They are without bloom. The flesh is rather thin and the stone is large and ovoid. The cherries are solitary or in small clusters. The shrub usually fruits abundantly. The cherries are sweet and edible. August.

Leaves. — The leaves are obovate-lanceolate, with a narrowed base. They are deep green above and paler beneath. The margin is toothed with the exception of a short distance

at the base, which is entire. The leaves change
to a deep red in autumn.

Flowers. — The white blossoms grow in few-
flowered clusters.

This trailing or prostrate shrub sends up erect
branches which are sometimes four feet high.
The plant sends out suckers freely, and spreading thus soon forms clumps. It grows in sandy
places along the eastern coast south to New
Jersey. It also occurs along the shores of the
Great Lakes.

WILD BLACK CHERRY. RUM CHERRY

Prunus serotina   **Plum Family**

Fruit. — The black drupes grow in loose
clusters at the ends of leafy branches. Many
of the flowers do not develop, and the cluster
often has a scraggly appearance. In ripening
the fruits change from green through yellowish
red, red, and dark red to black. The separate
cherries are spherical and flattened vertically.
A tiny depression is at the summit and the
persistent calyx is at the base. The pedicels
are short. The flesh is yellow or reddish and
rather thin. It is sweet, and although some-

WILD BLACK CHERRY (*Prunus serotina*)
199

what bitter has a pleasant flavor. August,
September.

Leaves. — The dark green glossy leaves, with
their whitened under surfaces, are alternate,
and usually ovate or oval-lanceolate. The apex
is pointed and the base rounded or narrowed.
The teeth are so incurved as to appear blunt.
The upper surface of the leaf stem is grooved
and bears two or more small glands near the
base of the leaf. Yellow is the autumnal color.

Flowers. — The small white flowers grow in
long, loose racemes. May, June.

The tree sometimes grows to a height of
eighty or ninety feet. The branches and bark
of the young trees are reddish brown. The
trunks of the older trees are almost black and
the bark is scaly. The wood is hard and has a
beautiful close grain. It is red and darkens
with age, does not shrink nor warp, and is
used for cabinet work and the inside finishings
of houses. It is becoming scarce. The cherries
are used in flavoring brandies and other intoxi-
cants. Both bark and fruit are ingredients in
certain medicines. The tree is common throughout the eastern part of the United States and
extends west to Dakota, Kansas, and Texas.

BLACK CROWBERRY

Empetrum nigrum   **Crowberry Family**

Fruit. — The black drupe is berrylike, globular, and incloses six to nine seedlike nutlets with a seed in each. The calyx is at the base and the stigma is at the apex. The drupes are solitary in the leaf axils. They are juicy, acid, edible, and serve as food for the Arctic birds.

Leaves. — The linear-oblong leaves roll their edges backward until they meet. They are dark green, thick, obtuse, and crowded along the branches. Evergreen.

Flowers. — The purplish diœcious flowers are small and solitary in the upper axils. The stamens are much exserted.

The Black Crowberry appears south to the coast of Maine, the higher mountains of New England, in northern New York, Michigan, and California. It is a native also of Europe and Asia. It is a much-branched shrub, low, densely leafy, and grows in thick beds.

INKBERRY. EVERGREEN WINTER BERRY

Ilex glabra   **Holly Family**

Fruit. — The berrylike black drupe is about a quarter of an inch in diameter. It is usually solitary in the leaf axils, with the calyx at the base and the mark of the stigma at the summit. The six seedlets are smooth.

Leaves. — The leathery evergreen leaves are wedge-lanceolate or oblong. The apex is obtuse or acute and sometimes few-toothed. The remainder of the margin is entire. The upper surface is dark green and shining and the lower one paler and dotted with black.

Flowers. — The small dioecious or perfect flowers grow in the leaf axils. They are borne on slender stems. June.

This slender evergreen shrub is from two to six feet high. It has long been cultivated in England, but with us occurs mainly in a wild state. It grows near the coast from Nova Scotia to Louisiana.

INKBERRY (*Ilex glabra*)

BUCKTHORN

Rhamnus cathartica   **Buckthorn Family**

Fruit.— The berrylike drupes grow in clusters. They are globose, somewhat flattened, black, and shining. The pulp and juice of the fruits are a peculiar green. The three or four inclosed nut- lets are grooved. The drupes are bitter and nauseating. August.

Leaves.— The leaves are broadly ovate with prominent, sometimes hairy, veins, beneath. They are finely toothed.

Flowers.— The small greenish flowers are diœcious. They appear a little later than the leaves. May, June.

This is a shrub or small tree ten or fifteen feet high. The lower branches, while leafy, are short and stiff and end in sharp points, really serving the purposes of thorns.

The berries were formerly used in medicines as a purgative, but are so severe in their action that their use in this direction is now confined to veterinary practice. A green dye is yielded by the ripe berries, a purple dye by the over-ripe fruit, a yellow dye by the fresh bark, and a brown one by the dried bark.

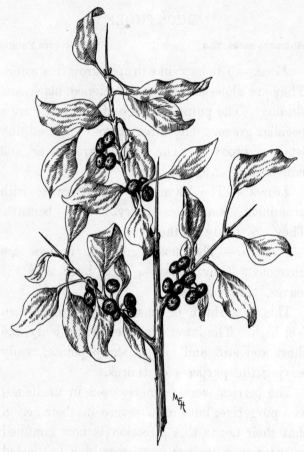

BUCKTHORN (*Rhamnus cathartica*)

The plant is an escape from hedges in New England and the Middle States, and was introduced originally from Europe. It is also a native of northern Asia.

LANCE–LEAVED BUCKTHORN

Rhamnus lanceolata   **Buckthorn Family**

Fruit. — This berrylike drupe has two grooved nutlets. It is black and shining. The fruits are in the leaf axils, sometimes in clusters of two or three.

Leaves. — The leaves are oblong-lanceolate. On the flowering shoots the leaf apex is often obtuse. The leaves are finely toothed and somewhat hairy on the veins beneath.

Flowers. — The yellowish green blossoms are solitary or clustered in the leaf axils. May.

The Lance-leaved Buckthorn grows along banks of streams and on hills from Pennsylvania southward. It is a tall, thornless shrub.

ALDER–LEAVED BUCKTHORN

Rhamnus alnifolia   **Buckthorn Family**

Fruit. — The black berrylike drupes are somewhat pear-shaped. They are fleshy and inclose three grooved nutlets. They grow from the axils of the lower leaves on shoots of new growth.

Leaves. — The broadly ovate leaves are dark green when fully grown. They are acute at the apex and the margin is bluntly toothed. The veins on the lower surface are very prominent.

Flowers. — The greenish flowers are mostly diœcious and grow on short stems in the leaf axils. May, June.

This is a low thornless shrub which grows in swamps, with New Jersey for its southern limit in our section.

NORTHERN FOX GRAPE

Vitis Labrusca   **Grape Family**

Fruit. — The fruit is very variable in color, in size of separate berries and of the cluster, and in flavor. Purplish black is the common color but ripe reddish and greenish fruits are found. The cluster is usually rather small. The berries are large, with a thick skin, tough pulp, and large, thick seeds. They drop readily when ripe. They are sweet, with a somewhat "musky" taste and odor. September.

Leaves. — The leaves are likewise variable; sometimes nearly entire in outline and broadly heart-shaped, sometimes three-lobed near the top, and sometimes five-lobed. The margins have sharply tipped teeth. The leaves are thick and large, with a green, nearly smooth upper surface and an under surface thickly covered with whitish or brownish wool.

Flowers. — The inconspicuous flowers are sometimes perfect, sometimes staminate or pistillate. The fertile racemes are compact.

This luxuriant vine climbs by means of its tendrils, which are modified flower peduncles, over rocks and walls, and from tree top to tree top in the forest. It is our most common grape and the one from which many cultivated forms, such as the Catawba, Concord, and Worden have sprung. Its berries are considered the best for jellies and are also valuable for grape juice. This delicious beverage is justly growing in favor. It is rich in nutriment, containing as much nitrogenous matter as milk.

The scurfy covering of the branches, stalks, and tendrils, together with the presence of tendril or flower cluster opposite each leaf, are distinguishing features. The bark peels off in shreds.

It occurs in New England and along the Alleghanies to central Georgia.

SUMMER GRAPE

Vitis æstivalis   **Grape Family**

Fruit. — The berries are medium-sized, one-third to one-half inch in diameter. They are dark blue or black with a bloom. The skin is tough; the flesh sometimes dry and puckery, sometimes sweet and juicy, always lacking the musky flavor of *Vitis labrusca;* and the seeds are small. The clusters are rather long, with long stems. September.

Leaves. — The large leaves, thickish when mature, are angled or three- to five-lobed. The openings between the lobes are deep or broad and open. The base is heart-shaped. The young leaves are shining above and have tufts of brown down on the lower surfaces. The older leaves are dull green above and with the distinguishing brown woolly tufts along the veins.

Flowers. — The flower cluster is long and loose.

This vine, like the preceding, is one of vigorous growth, and has given rise to several cultivated varieties. It is distinguished by its brown woolly masses on the leaves and by the absence of tendril or inflorescence opposite each third leaf.

Bailey, in his " Evolution of our Native Fruits," gives its range as Chenung County, New York, and Long Island to central Florida and west through southern Pennsylvania to the Mississippi and Missouri.

It is especially a southern grape, whose place in the north is represented by the next plant, *Vitis bicolor*, which is considered by Gray a variety of *Vitis æstivalis* but as a separate species by Bailey and by Britton and Brown.

BLUE GRAPE

Vitis bicolor   **Grape Family**

Fruit. — The clusters are usually long, with a long peduncle. The berries are purple, with a dense bloom, medium in size, and sour in taste. The seeds are small. September.

Leaves. — The large, usually three- to five-lobed leaves have not as deeply notched teeth as those of· *Vitis æstivalis.* A distinguishing feature is the thick blue bloom on the under surface of the leaf. It loses this toward fall, but does not have the brown woolly masses of the Summer Grape. The petioles and tendrils are long.

The young growths, as well as the under leaf surface, are usually covered with the distinguishing blue bloom. The vine grows along streams and on banks from New York to Illinois and to mountains of North Carolina and Tennessee.

RIVERSIDE OR SWEET-SCENTED GRAPE

Vitis vulpina.   Vitis riparia   **Grape Family**

This species differs from *Vitis cordifolia* (the following species) chiefly in the following particulars : —

Fruit. — The berries are thickly covered with blue bloom. The seeds are small. The fruit clusters are much-branched and often compound.

Leaves. — These show deeper and more frequent lobes. The veins and angles are often hairy.

Flowers. — The blossoms are very fragrant. They grow in smaller, denser clusters.

This has a range from New Brunswick to

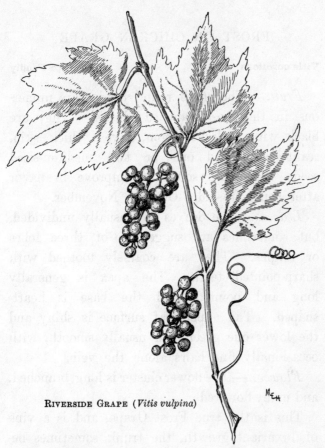

RIVERSIDE GRAPE (*Vitis vulpina*)

North Dakota, Kansas, and Colorado, south to
West Virginia, Mississippi, and Texas. It is the
source of some cultivated species; Elvira, Clinton,
and others.

FROST OR CHICKEN GRAPE

Vitis cordifolia   **Grape Family**

Fruit. — The small round berries are numerous in the loose-branched cluster. They are black with a slight bloom, have a thick skin, scant pulp, and one or two medium-sized seeds. They are sour, but improve in flavor after being frosted. October, November.

Leaves. — The leaves are usually undivided, but sometimes are suggestive of three lobes or angles. They are coarsely toothed with sharp-pointed teeth. The apex is generally long and pointed, and the base is heart-shaped. The upper leaf surface is shiny and the lower one green and usually smooth, with occasionally fine hairs along the veins.

Flowers. — The flower cluster is long, branched, and many-flowered.

This is the true Frost Grape, and is a vine of luxuriant growth, the trunk sometimes becoming a foot or two in diameter. It grows in moist thickets and along streams from New England to central Illinois, Missouri, Nebraska, and southward.

VIRGINIA CREEPER (*Parthenocissus quinquefolia*)

216

VIRGINIA CREEPER. WOODBINE
AMERICAN IVY

Parthenocissus quinquefolia. Ampelopsis quinquefolia
Grape Family

Fruit. — The globular berry is slightly depressed at the tip. It is dark blue or nearly black when mature, is two-celled, with one or two largish seeds in each cell. The flesh is thin and inedible. The berries grow in loose red-stalked clusters. October.

Leaves. — The compound leaves are borne on long channeled red stems. The five to seven leaflets are in a whorl at the apex of the leaf stalk. They are variable in shape, oval or elliptical, and are coarsely toothed along the apex half of the margin. The stems are short and the apex is long and acute. The leaves early assume their red, crimson, and purplish fall colorings.

Flowers. — The reddish or greenish small flowers grow in cymes. Despite their inconspicuous appearance and apparent lack of odor, they are visited by many bees.

Virginia Creeper is a vine which has been

much cultivated. It grows rapidly, and covers house walls and various supports offered it. When growing wild, it climbs tree trunks and covers stone walls, fences, and rocks. It supports itself by means of the small disks at the ends of the tendrils. The fall coloring is brilliant. Its five leaflets are a feature distinguishing it from the Poison Ivy, which has somewhat similar habits of growth. The latter's leaflets are but three in number.

ANGELICA TREE.   HERCULES' CLUB

Aralia spinosa   **Ginseng Family**

Fruit. — The black berrylike drupes are five-lobed and bear the styles at the summit. The fruits grow in large terminal clusters. The flesh is thin. The fruits hang on the trees during the winter.

Leaves. — The leaves are doubly or triply compound and very large. The leaflets are ovate, thick, and serrate. They are dark green above and paler beneath. The petioles are prickly. Dark red with traces of yellow is the fall coloring.

Flowers. — The small white flowers grow in umbels, some of which form a large panicle.

In the south, this plant is said to become a tree of fifty feet in height. In our section, however, it is a small tree or large shrub. The branchless stems often grow in groups, bearing their large compound leaves in clusters at the top. The general effect is somewhat like a palm. The stems and leaves are thorny. The flowers, like those of the Spikenard, are late in appearing and the fruit matures rapidly. Southern New York is the northern limit, although it is often cultivated farther north and sometimes escapes.

AMERICAN SPIKENARD. INDIAN ROOT

Aralia racemosa   **Ginseng Family**

Fruit. — The large raceme-like cluster of fruits is composed of numerous umbels. Smaller clusters grow in the leaf axils. The berry is small, gobular, dark purple or reddish brown, five-seeded, and crowned with tiny calyx teeth, through which the styles project. The berries, like the roots, are aromatic. September.

Leaves. — The leaves are large and compound.

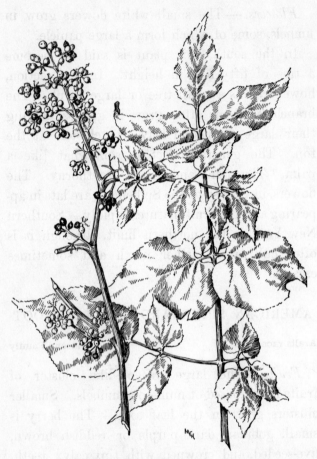

AMERICAN SPIKENARD (*Aralia racemosa*)

Each has a main stem and two opposite lateral
branches. Three to five leaflets, a terminal one

and the others in pairs, grow on each of the stems. The leaflets are sometimes lobed. They are usually heart-shaped and sharply and doubly toothed. The point is long and sharp, and the base is heart-shaped. The veins on the lower surface are hairy.

Flowers. — The small, greenish, umbelled flowers form long terminal spikes or smaller spikes in the leaf axils. July, August.

Along the wooded roadsides, the greenish white flowers appear about the time that the Golden-rod begins to blossom. The fruit follows in haste; and the plant, with its tiny glassy spheres, is more noticeable than in its period of bloom. The berries are used as food by birds.

Its range is from New Brunswick to Georgia and west to Minnesota.

WILD OR VIRGINIAN SARSAPARILLA

Aralia nudicaulis   **Ginseng Family**

Fruit. — The fruit is borne on a naked scape, which is shorter than the leaf stalks. There is usually one large cluster of fruits at the top of the scape. This cluster is often composed of

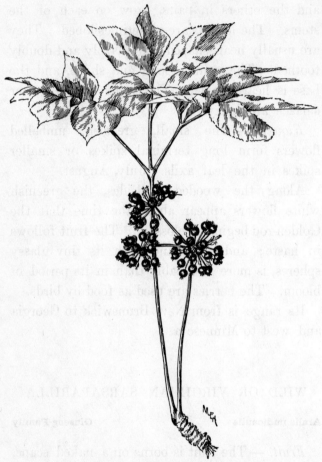

WILD SARSAPARILLA (*Aralia nudicaulis*)

two small clusters borne on short stems. From
this central cluster radiate, on longer stems, one

or more smaller clusters of fruits. The clusters are compact and globular in appearance. The black berrylike drupes composing them are globular and about the size of peas. At the top of each is visible the opening of the calyx tube, with its minute teeth. Projecting through and beyond this are the five styles. The drupe is five-celled, with one nutlet in each cell. The green fruits are ridged, showing the five-celled structure externally; but when ripe the drupes are nearly smooth. July, August.

Leaves.—There is usually one, sometimes there are two, long stalked, compound leaves. Each leaf has three divisions of five to seven leaflets each. These are finely toothed and acute at the apex.

Flowers. — The flowers are greenish white, and are borne in umbels composed of from three to seven clusters of bloom.

The aromatic root serves as a substitute for the South American Sarsaparilla. Bluebirds are recorded as eating the fruit. It favors damp woods, and extends south from Newfoundland to North Carolina and west to the Dakotas.

BRISTLY SARSAPARILLA. WILD ELDER

Aralia hispida   **Ginseng Family**

Fruit. — The dark blue, almost black, berry-like drupes are usually five-seeded. When the fruit is green, or the berry somewhat dry, it shows its five parts very distinctly. The five styles protrude through the persistent calyx tube at the top of the fruit. Several umbels on very slender smooth pedicels grow at the summit of the plant stem. August.

Leaves. — The leaves are twice pinnate, with long ovate leaflets. These are finely toothed, sharply acute at the apex, narrowed or rounded at the base, and hairy on the veins beneath.

Flowers. — The tiny white flowers grow in nearly hemispherical clusters.

This Sarsaparilla is distinguished by the bristles which are scattered along the stem. It grows from one to two feet high, and frequents rocky and sandy places. It extends south to North Carolina.

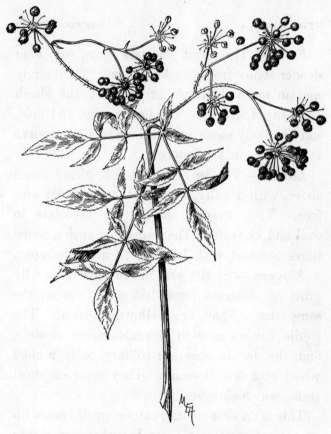

BRISTLY SARSAPARILLA (*Aralia hispida*)

TUPELO.   SOUR GUM.   PEPPERIDGE

Nyssa sylvatica   **Dogwood Family**

Fruit. — The fruit clusters grow on long
slender stems from the leaf axils.   They rarely
contain more than two or three of the bluish
black ovoid drupes.   The flesh is thin and acid,
and the bony stone grooved.   The drupes serve
as food for birds.   October.

Leaves. — The leaves are a soft glossy green
above, with a paler, somewhat hairy under surface.   They vary in shape from lanceolate to
oval and obovate.   They are often entire, sometimes notched, with large teeth near the apex.

Flowers. — Sterile and fertile flowers usually
grow on different trees, but sometimes on the
same tree.   They are yellowish green.   The
sterile flowers grow in several-flowered clusters,
and the fertile ones are solitary, or in a close
whorl of a few blossoms.   They grow on short
stalks which elongate in fruit.

This is an ornamental, rather small tree, with
an attractive foliage.   Its branches are rather
low, horizontal, and quite close.   The wood splits
with difficulty on account of its twisted fibers.

Tupelo (*Nyssa sylvatica*)

227

Tupelo is the Indian name for the tree. An interesting tradition in connection with the tree still clings around slavery days. It was customary to use a log of the Sour Gum as the back log for the rousing Christmas fire. As long as the fire lasted, work on the plantation was suspended. Prompted by the characteristic love of leisure possessed by the colored race, the slaves would cut a large log in the fall, sink it under water, and leave it there until near Christmas, when they would raise it and carry it in with the other Christmas fuel. Full of water, it burnt a long time, and the slaves enjoyed a correspondingly long vacation. The tree ranges from New England west to Michigan, and south to Florida and Texas.

ALPINE OR BLACK BEARBERRY

Mairania alpina   **Arctostaphylos alpina**
                Heath Family

Fruit. — The globose drupes are black and juicy. They inclose four or five separate nutlets, each one-seeded. They grow in small clusters.

Leaves. — The leaves are deciduous, toothed, and inversely egg-shaped, with conspicuous veinings.

Flowers. — The white ovoid flowers have a narrow throat. They grow in terminal racemes.

This Bearberry is an Arctic mountainous shrub around the world. It also occurs in the mountains of New England and Canada. It is depressed, not half a foot high.

BLACK OR HIGH-BUSH HUCKLEBERRY

Gaylussacia resinosa   **Huckleberry Family**

Fruit. — The black, shining, berrylike drupes grow in short racemose clusters. The calyx teeth are plainly visible. A cross section of the fruit near the base shows the circular arrangement of the ten nutlets around the core. This core tapers toward the summit, being somewhat cone-shaped. July, August.

Leaves. — The thick green leaves are covered with resinous dots. They are entire and have short petioles. They vary from oblong to oval, and are obtuse or acutish. A purplish red is one of the most noticeable of its fall colorings.

Flowers. — The reddish or pink bells grow in short one-sided racemes.

There are several varieties with berries differing from the type; some, pear-shaped; some, bluish; and some, black with a bloom.

Gaylussacia resinosa is the Huckleberry commonly for sale. The flesh is harder than that of the Blueberries, but the hard nutlets are somewhat objectionable.

Huckleberries and milk! What recollections of childhood the combination recalls! Bluebirds, robins, cedar birds, crows, and blue jays share with mortals a liking for the berries.

The Huckleberries contribute an important share to the beauty of the autumnal display of colors. Great purplish patches on pasture hillsides are visible for a considerable distance.

The species extends as far south as Georgia, and west to Minnesota.

DWARF OR BUSH HUCKLEBERRY

Gaylussacia dumosa   **Huckleberry Family**

Fruit. — The berrylike drupes, with their ten seedlike nutlets, are small, watery, and insipid.

They are black and shining, without bloom, and grow in an open bracted cluster. July, August.

Leaves. — The thick leaves are green on both sides, shining when old, and resinous. They are nearly or quite stemless, and often slightly downy. The apex is obtuse or acute, and ends in a sharp point.

Flowers. — The racemes of white, pink, or red bell-shaped flowers grow in loose bracted racemes. June.

The fruit of this variety is not of much account. The plant is the principal member of the genus southward. It grows in sandy swamps along the coast from Newfoundland to Florida and Louisiana.

LOW BLACK BLUEBERRY

Vaccinium nigrum　　　　　　　　　**Huckleberry Family**

Fruit. — The berry is black and has no bloom. July.

Leaves. — The leaves are oblong, obovate, or oblanceolate. They are nearly sessile and finely toothed. The apex is acute and the base narrowed or rounded. The under surface is pale and whitened.

Flowers. — The flowers are white and bell-shaped. The bell is rounder than the blossom of *V. Pennsylvanicum.* Only a few flowers appear in the cluster.

This is sometimes considered as a variety of *V. Pennsylvanicum,* and often grows with it. It differs from it in having a rounder, bell-like blossom and in the black bloomless fruit.

Vaccinium atrococcum is sometimes considered as a variety of *V. corymbosum,* which is described in the blue section. The stems and under leaf surfaces are downy. The berries are black and lack bloom.

FRINGE TREE

Chionanthus Virginica   **Olive Family**

Fruit. — The purple oval drupes grow in loose clusters. They are covered with a bloom. The four-parted calyx is persistent at the base and the style is at the tip. The dry flesh contains one stony seed. The skin is thick.

Leaves. — The ovate or obovate-lanceolate leaves have stout, hairy stems. They are entire, and sharp or rounded at the apex. The under surface is hairy along the veins. The leaves turn yellow in the fall.

Flowers. — The white flower clusters are decorative among the green foliage. The four narrow petals, hanging like fringes, give the common name to the plant.

The Greek, *Chionanthus*, meaning snow and blossom, refers to the white flowers. This shrub, or small tree, is native as far north as New Jersey and southern Pennsylvania, and extends southward to Florida and west to Texas, Arkansas, and Kansas. It grows along the banks of streams. It is often cultivated at the north.

PRIVET

Ligustrum vulgare   **Olive Family**

Fruit. — The shining black berries are from one- to two-seeded. They grow in terminal panicles.

Leaves. — The leaves are deciduous with us, but in the south of Europe are evergreen. They are entire and very smooth.

Flowers. — The small white flowers are in terminal panicles.

The leaves and bark are astringent. In

Belgium and in other parts of Europe, the small twigs are powdered and used for tanning leather. The juice of the berries is used in dyeing. This is a hardy shrub from six to eight feet high. It has been naturalized from Europe. It is often used for hedges. Some of its old English names are Primwort, Skedge, and Skedgwith.

Privet is reported growing on the walls of Cologne Cathedral, the seeds obviously having been deposited there by bird agencies.

BLACK OR GARDEN NIGHTSHADE

Solanum nigrum   **Potato Family**

Fruit. — Smallish, black, globular berries grow in drooping clusters from the side of the stems. Their pedicels are slender, and the five-parted calyx is at the base. The berries are smooth and contain many thin, flat seeds.

Leaves. — The ovate leaves usually have one side which is slightly longer than the other. They are wavy-toothed, thin, and have thin stems.

Flowers. — The five-lobed white flowers grow in lateral clusters. July–September.

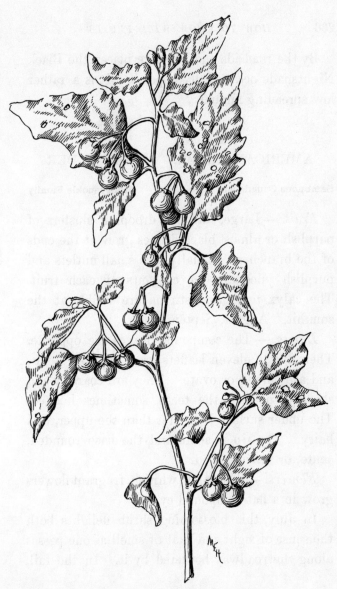

BLACK NIGHTSHADE (*Solanum nigrum*)

By the roadside and in waste places, the Black Nightshade occasionally appears. It is a rather low spreading annual.

AMERICAN ELDER.   SWEET ELDER

Sambucus Canadensis   **Honeysuckle Family**

Fruit. — Large, full, flat, drooping clusters of purplish or almost black drupes grow at the ends of the branches. Usually, five small nutlets and purplish juice are the contents of each fruit. The calyx teeth and stigma are visible at the summit. August, September.

Leaves. — The compound leaves are opposite. Their five to eleven leaflets grow on short stems, and are oblong or ovate. They are coarsely and sharply toothed, the teeth sometimes hooked. The under surface is lighter than the upper, and hairy. The tip is acute and the base rounded, acute, or heart-shaped.

Flowers. — The small, whitish, fragrant flowers grow in a flat compound cyme.

In July, this blossoming shrub delights both the sense of sight and that of smell as one passes along the roadway bordered by it. In the fall,

at school-opening season, the drooping clusters
of fruit are a feast for the eye; are sometimes
used for pies and homemade wine; and furnish
material for the country boy's ink bottle, much to
the distress of his schoolma'am. Professor Budd
is responsible for the statement, that with the
addition of an acid, vinegar or lemon juice,
Elderberries make as good a pie as Huckleberries.

The new growths are smooth and green, and
the older stems are grayish, with raised dots.
The pith is white, distinguishing this Elder from
the Red-berried, which has a brown pith. It is a
common plant of the United States and Canada.

MAPLE–LEAVED VIBURNUM OR ARROW-
WOOD. DOCKMACKIE

Viburnum acerifolium   **Honeysuckle Family**

Fruit. — The smallish drupes are somewhat
oval in shape, with two opposite sides flattened.
They are pointed at the tip. They are nearly
black when ripe. The flesh is thin and the
stone is doubly convex, with one ridged surface
and the other one slightly two-grooved. The

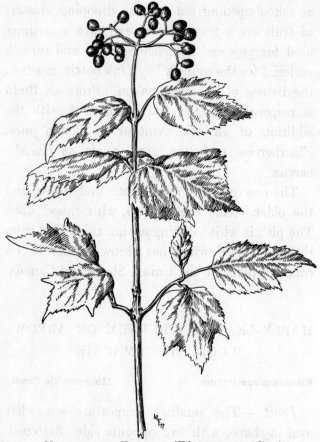

MAPLE-LEAVED VIBURNUM (*Viburnum acerifolium*)

fruits are borne in a terminal flat cluster, on reddish downy stems. Late August, September, and persistent through the winter.

Leaves. — The leaves are in pairs, with tiny stipules at the base of the stems. The under surface of the leaf is lighter than the upper, and is soft, with down. The shape varies from oval to somewhat three-lobed. The leaves are unevenly toothed.

Flowers. — The flower cluster consists of perfect, small, white, or cream-colored blossoms. May, June.

This is a low shrub, seldom exceeding six feet in height. It is quite readily distinguished by the resemblance of its leaves to those of the Red Maple. It grows on the border of woods south to North Carolina and west to Michigan and Minnesota.

DOWNY-LEAVED ARROWWOOD

Viburnum pubescens   **Honeysuckle Family**

Fruit. — The dark purple oval drupes are clustered. The flesh is thin and the stone is two-grooved on each surface. August.

Leaves. — The ovate leaves are stemless or nearly so. They are coarsely toothed and acute at apex, or sometimes the point is long drawn

out.   The under surface is soft with down.
Purple and red are the foliage colors of autumn,
which contrast with the dark berries.

Flowers. — The abundant white flowers grow
in an open cyme.

The sessile, or nearly sessile, leaves and their
soft pubescence are characteristics of this low
branching shrub of rocky woods.   It extends
south along the Alleghanies to Georgia.   It
ranges west to Minnesota and Iowa.

WITHE–ROD

Viburnum cassinoides   **Honeysuckle Family**

Fruit. — The globose or ovoid drupes are
borne on red stems.   The cluster presents a
most attractive appearance with light green,
pink, and blue-black fruits in various stages of
ripening.   The dark drupes are covered with a
soft blue bloom.   The minute calyx and stigma
persist at the tip.   The flesh is quite abundant
and sweet.   The stone is flat, with a slight hollow on one side and a convex surface on the
other.   Late August, September.

Leaves. — The thickish opposite leaves grow
on flattened petioles, which nearly encircle the

WITHE-ROD (*Viburnum cassinoides*)

241

smaller branches. Brown circular dots appear on the upper surface of the leaf along the midvein, and are scattered about on the under surface. The leaf is usually ovate, with a blunt tip. The teeth are fine and somewhat rounded, or the margin is sometimes entire.

Flowers. — The flower cluster is quite large and full. The whitish flowers are small, perfect, and five-parted. June.

The shrub is rather straggling, and has an ash-colored bark. The twigs are somewhat scurfy and dotted. The slender last year's growth is sometimes used in binding sheaves. It is a swamp plant, and extends south to New Jersey and west to Minnesota.

LARGER WITHE–ROD

Viburnum nudum   **Honeysuckle Family**

This is usually a larger species than *Viburnum cassinoides*, and has a southern range extending from New Jersey south to Florida. The leaves are more prominently veined than in the preceding, and sometimes scurfy above. The margin is generally entire.

SWEET VIBURNUM. SHEEPBERRY
NANNY BERRY

Viburnum Lentago   **Honeysuckle Family**

Fruit. — The drupes are crimson, before ripening to a dark blue or black, and the two colors often mingle in the fruit cluster. The fruits are covered with bloom, are drooping, and the clusters have slender red stalks. The calyx tube, with the projecting stigma, is at the summit. The fruit is rather large and edible. The stone is flattened, has a blunt point, and is grooved on both sides. September, October.

Leaves. — The broad oval leaves are sharp-pointed, and are sharply and closely toothed. The leaf stem is usually winged or margined. In the fall the leaves are deep red or marked with orange.

Flowers. — The small white flowers grow in terminal cymes. The numerous yellow anthers give the flower a yellowish appearance. May, June.

This small tree has rusty, scurfy, scale-like bark, especially on the young shoots. Its foliage is good, and the flower clusters large and showy.

SWEET VIBURNUM (*Viburnum Lentago*)

245

It occurs quite frequently in woods and along streams from Canada to Georgia, and west to Minnesota and Missouri.

BLACK HAW. STAG BUSH

Viburnum prunifolium   **Honeysuckle Family**

Fruit. — The dark blue, nearly black, oval drupes are borne in a few-fruited cluster. The fruits are whitened with a bloom. The oval stone is flat on one side and a trifle curved on the other. The flavor of the fruit is improved after having been frosted. September.

Leaves. — The oval leaves are usually obtuse at the apex and finely toothed. They are dark green above and lighter beneath. They grow on short stems which are sometimes winged.

Flowers. — The cream-white flowers grow in a flat-topped cluster. May.

This Viburnum, like the Sweet Viburnum, sometimes reaches the stature of a tree. It is found from Connecticut to Florida, and extends west to Michigan, Kansas, and Texas.

BLUE

BLUE

COMMON JUNIPER

Juniperus communis   **Pine Family**

Fruit. — The berrylike cones do not develop
until the second year, and often remain on the
branches some time after ripening. When fully
ripe, in the fall of the second year, the fruits
are dark blue with a bloom. They are usually
three-seeded. The flesh of the berry is dry and
mealy. The seeds are slow in germinating, re-
quiring two years. The fruit develops from
three fleshy scales, united from their bases nearly
to the tips, and inclosing three ovules. When
ripe, the tips of the scales are still visible, with
lines from each joining in a common center.
The berry is nearly stemless and axillary. It
is much used in making gin, an infusion of the
berries being added to distilled grain.

Leaves. — The short, stemless, sharp-pointed
leaves are arranged in whorls of three. They
are bright green and shining on the lower sur-

Low Juniper (*Juniperus nana*)

250

face, and channeled and whitened on the upper one. The whitened appearance of the upper surface is due to a thin layer of wax, which covers and protects, from dew and rain, the stomata, or openings, of the air passages.

Flowers. — The staminate and pistillate flowers grow in aments on separate plants. April, May.

Juniperus communis is an erect shrub or small tree, common to the northern portions of Europe, Asia, and America. In the latter continent it extends as far south as New Jersey, Pennsylvania, Nebraska, Michigan, and along the Rockies to New Mexico.

Juniperus nana (*Juniperus communis*, var. *alpina* of Gray) is distinguished from the preceding by a growth in low circular patches. These spread over waste rocky hillsides and are eradicated with difficulty. The leaves are somewhat stouter and less spreading than those of *Juniperus communis*.

RED CEDAR

Juniperus Virginiana   **Pine Family**

Fruit. — The fruits are globular or flattened
and wider at the top, giving the " berry " a
triangular outline. The so-called " berry " is
formed by the coalescence of fleshy scales, the
tips of which are indicated by tiny projections
on the fruit. It grows on a straight peduncle
and contains one or two seeds. Seeds and
flesh are aromatic. October, November, and
persistent.

Leaves. — The leaves are of two kinds. On
the younger trees they are often awl-shaped and
arranged loosely along the branches. These
also appear on older trees, together with short,
scale-like, overlapping leaves crowded closely
together.

Flowers. — The sterile and fertile flowers are
usually on different trees, sometimes on the
same tree. The flowers are small and grow in
terminal aments. April, May.

The Red Cedar is a shrub or tree with reddish
brown bark, which peels off in shreds on the
older growths. The wood is whitish or red, and

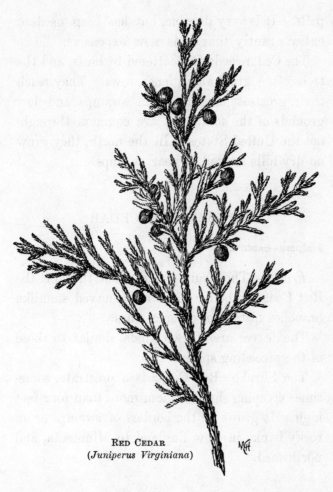

RED CEDAR
(*Juniperus Virginiana*)

has a pleasant, persistent odor. It is used for
pencils, small boxes, fence posts, and sometimes

pails. It is very durable, but has been used so extravagantly that it is now expensive.

The Cedar seeds are scattered by birds, and the trees often grow along fence rows. They reach their greatest magnitude in swamps and low grounds of the south, but are common throughout the United States. In the north, they grow on dry hills as well as near swamps.

SHRUBBY RED CEDAR

Juniperus Sabina　　　　　　　　　　　　**Pine Family**

Fruit. — The fruit differs from that of the Red Cedar in being borne on recurved stemlike branches instead of on erect ones.

The leaves are of two kinds, similar to those of the preceding species.

The Shrubby Red Cedar is a prostrate, sometimes creeping shrub, seldom more than four feet high. It grows on the borders of swamps or on rocky banks in New England to Minnesota, and northward.

Yellow Clintonia (*Clintonia borealis*)

256

YELLOW CLINTONIA

Clintonia borealis   **Lily-of-the-Valley Family**

Fruit. — The ovoid berry is almost a pure blue in color. It is many-seeded. The umbel of fruit grows at the top of an erect stem. August.

Leaves. — There are two to four shiny, oval or oblong, light green leaves, with their stalks acting as a sheath for the base of the scape.

Flowers. — The three to six, greenish, drooping flowers grow at the summit of the scape. May, June.

The plant is named in honor of De Witt Clinton, who was a governor of New York State and a naturalist. It grows from six inches to a foot in height. The rootstock is slender and creeping.

It is found in woods from Labrador to North Carolina. Its western limit is Minnesota.

BLUE COHOSH. PAPOOSE ROOT

Caulophyllum thalictroides   **Barberry Family**

Fruit. — The fruit resembles a drupe, but is a naked seed with the outer coat fleshy. There are originally two seeds in the developing ovary. As these grow they burst their membranous covering and continue growth as pairs of naked seeds. The fruit is blue with a bloom, globular, and borne on short stout stalks. The fruits grow in raceme-like clusters.

Leaves. — There is one large leaf at the top of the stem and sometimes a smaller one near the base of the flower. The compound leaf is thrice-parted and the leaflets have two or three lobes. They are coarsely toothed.

Flowers. — The flowers are yellowish green, small, and in racemes. April, May.

This herb of early growth appears, in rich woods, in April. When young, the whole plant is bloom-covered. It is more common to the westward, and extends as far south as South Carolina.

SASSAFRAS

Sassafras sassafras   **Sassafras officinale**
Laurel Family

Fruit. — The fruit is an oval dark blue drupe. This fits into a red hollow cup, which is thickened calyx and fleshy stem. The calyx teeth

BLUE COHOSH (*Caulophyllum thalictroides*)
259

scallop the edge of the cup. The flesh of the drupe is rather thin and the stone large. The cotyledons are large and fleshy. The fruits grow singly or in small clusters from the base of the season's shoots. The fruit is eaten by birds, but is unpleasantly spicy. August.

Leaves. — On the mature trees, oval leaves predominate. The young shoots bear oval leaves; leaves with a lobe at one side, looking like the thumb of a mitten; or three-lobed leaves, with two lateral lobes and a terminal one. The hollows of the lobed leaves are rounded. The young leaves are reddish but become dark green above with a lighter lower surface. The leaves and twigs are mucilaginous. Yellow and orange are the fall colors.

Flowers. — The greenish yellow diœcious flowers grow in drooping many-flowered racemes.

Sassafras and Spice Bush are our only representatives of a large family that, in the tropics, include plants that yield cinnamon, camphor, and several differently scented woods. The Laurel or Bay Tree, whose leaves were used by the ancients in making wreaths with which to crown their heroes, is also a member of this family.

Aromatic bark is a common characteristic.
The bark and roots are ingredients in root beer,
and from the bark of the roots oil of sassafras is
made. The bark of the Sassafras is much cracked
and roughened. Emerson says that, in the south-
western parts of the country, the dried leaves
of the Sassafras are much used for flavoring
soups. Columbus is said to have increased his
own hope of being near land, and to have
quieted the mutinies of his crew, from catching
whiffs of the strong fragrance of the Sassafras.
Sassafras roots were a part of the first cargo to
be sent from Massachusetts to England. At
that time they were much prized for supposed
medicinal properties. The wood is brittle, but
when seasoned is tough and light. The trees
grow rapidly and spread by suckers, often forming thickets. The range is through the Mississippi valley and eastward.

ROUND-LEAVED CORNEL OR DOGWOOD

Cornus circinata   **Dogwood Family**

Fruit. — The small drupe is very light blue
or white. The fruit develops sparingly and the

cymes are not very full. The stone is nearly globose and somewhat ridged. It is aromatic and bitter. September.

Leaves. — The leaves are nearly round, sometimes even broader than long. The apex is acute and the base rounded or heart-shaped. The under surface is densely hairy and has prominent veins.

Flowers. — The white blossoms are rather large and in full-blossomed cymes. The pedicels are somewhat hairy.

This shrub is quite spreading in its habit, and from three to ten feet high. Its branches are green and warty. The leaves are distinctively broad. It grows in the shade and often among rocks. It extends from Nova Scotia to Virginia.

SILKY CORNEL. KINNIKINNIK

Cornus Amonum   **Cornus sericea**

Dogwood Family

Fruit. — The drupes vary in ripening from green to pale blue. They are globular, with the calyx teeth persistent in a depression at the summit. The flesh is whitish and the stone noticeably ridged. The fruits grow in a flat

SILKY CORNEL (*Cornus Amonum*)

terminal cluster. The peduncles and pedicels are reddish and clothed with soft down. Late August, September.

Leaves. — The simple, opposite leaves are ovate or elliptical. The tip is pointed and the base is rounded or often uneven, one side being longer than the other. The stems and under leaf surfaces are downy, sometimes rusty.

Flowers. — The small white flowers grow in flat compact cymes. June.

This shrub is erect and somewhat spreading. Its green bark has a reddish tinge and in winter the branches become purplish. The branchlets, stems, and lower leaf surfaces are finely woolly. It is one of the latest of the family to blossom but fruits in company with the Panicled Cornel, the two often forming hedges along the fence rows and highways. It is very decorative in fruit, and is being more and more used by landscape gardeners. It grows quite extensively as far west as the Dakotas and south to the gulf.

ALTERNATE–LEAVED CORNEL OR DOGWOOD

Cornus alternifolia   **Dogwood Family**

Fruit. — The small deep blue drupes grow in an irregularly branched drooping cyme. Peduncles and pedicels are a deep red. The flesh of the drupe is scanty, white or pinkish, and of a pithy texture. There is but one stone, which is globose and usually two-seeded. The style projects through the minute calyx tube, at the summit of the fruit. The drupe is tenaciously bitter. It ripens in early August, being one of the first Dogwoods to fruit.

Leaves. — The alternate leaves usually grow in clusters at the ends of the branches. They are entire or minutely toothed, and ovate or oval. The pointed apex is long drawn out and the base is rounded or acute. The upper surface is shining and dark green ; the lower one, whitened and covered with fine hairs, especially along the veins. The veins are prominent on the under surface, looking like tiny cords running through the leaf. The petiole has a grooved upper surface. Yellow or yellow and scarlet are the fall colors.

ALTERNATE-LEAVED CORNEL (*Cornus alternifolia*)

Flowers. — The small, white, four-parted flowers grow in broad loose cymes. June.

This is a pretty shrub or small tree, distinguished from the other Dogwoods by its alternate leaves. Its flower clusters, too, differ, the secondary stalks growing alternately instead of starting from the same point. The leaf clusters are broad and flat, and so arranged as to form a green background for the red and black of the fruit cluster. This Dogwood fruit, bitter though it is, serves as food for the birds. It is common from New Brunswick to Minnesota and as far south as Georgia.

BLUE TANGLE. TANGLEBERRY. DANGLE–BERRY

Gaylussacia frondosa **Huckleberry Family**

Fruit. — The long loose clusters of berrylike drupes are characteristic of this species. The separate fruits are rather large, dark blue with a white bloom, globose, and sweet with a slight acidity. The calyx teeth crown the summit. The fruits ripen late and are rather scarce. July, August.

Leaves. — The short-petioled leaves are thin, large, pale green, whitened and resinous on the under surface, and oval to inversely egg-shaped.

Flowers. — The greenish pink bells, as Emerson says, " hang dangling on slender strings, from one to three inches long." These stems are bracted, and the raceme-like flower cluster is long and loose. May, June.

In New England, the Dangleberry grows mostly along the coast. It extends south to Florida and Louisiana and west to Ohio. It prefers moist ground, and the fruit in the warmer locations is of improved quality.

BOX HUCKLEBERRY

Gaylussacia brachycera   **Huckleberry Family**

Fruit. — The light blue, berrylike drupes grow in short clusters. They have ten seedlike nutlets.

Leaves. — The evergreen leaves are thick and leathery, and lack the resinous dots common to the rest of the genus. They are oval, and the margins have rounded teeth and are somewhat rolled backwards. The leaf stems are very short.

Flowers. — The white or pink bell-shaped flowers grow on very short pedicels, in short racemes. May.

This low shrub, scarcely exceeding a foot in height, has a limited range, occurring in dry woods from Delaware and Pennsylvania to Virginia.

GREAT BILBERRY

Vaccinium uliginosum   **Huckleberry Family**

Fruit. — The blue, bloom-covered, globular berries usually grow singly or in clusters of from two to four. They are four- or five-celled, sweet, and not very abundant. July, August.

Leaves. — The oblong or obovate leaves, when fully grown, are thick, bright green above and paler beneath. They are entire and nearly stemless.

Flowers. — The solitary or few-clustered pink flowers have their parts mostly in fours.

This is a low tufted shrub with many branches. It inhabits the mountain heights of New England and New York, the shore of Lake Superior, and thence northward to Alaska. It is also found in the northern countries of the Eastern Hemisphere.

DWARF BILBERRY

Vaccinium cæspitosum   **Huckleberry Family**

Fruit. — The berry is globular and blue, with a bloom. It has a sweet flavor. It is five-celled. The fruits usually grow singly in the leaf axils. August.

Leaves. — The smooth shining leaves are obovate, with small blunt teeth. The petioles are very short.

Flowers. — The white or pink flowers are bell-shaped.

This is mainly a mountain or cold country shrub, and grows to a height of from four inches to two feet.

HIGH-BUSH OR TALL BLUEBERRY

Vaccinium corymbosum   **Huckleberry Family**

Fruit. — The berries differ much in color, some varieties bearing shiny black berries, some black with a blue bloom, and some blue. The

HIGH-BUSH BLUEBERRY (*Vaccinium corymbosum*)

size of the berry is also variable. The berries
grow in a cluster at the end of a short, nearly

leafless branch of last year's growth. The calyx teeth are noticeable at the summit of the berry. Some berries are very sweet and others rather acid. July, August.

Leaves. — In the typical form the margins are entire. After the time of flowering, the leaves broaden without increasing in length. They are oval or elliptical-lanceolate. The petioles are short. The under surface is paler than the upper and may be smooth or hairy.

Flowers. — The blossoms are white or pink-ish, cylindrical, and somewhat narrowed at the throat. They grow in short racemes.

The High Blueberry grows to a height of from four to ten feet. It forms a bushy shrub. On the older branches the bark roughens and comes off in shreads. The leaves add their scarlet and orange colorings to the brilliancy of the autumnal swamp foliage. These berries grow as far north as Newfoundland, west to Minnesota, and south to Virginia. While reach-ing their most luxuriant growth in swamps, they are also abundantly found in old pastures.

LOW BLUEBERRY (*Vaccinium vacillans*)

DWARF BLUEBERRY (*Vaccinium Pennsylvanicum*)

DWARF BLUEBERRY

Vaccinium Pennsylvanicum   **Huckleberry Family**

Fruit. — The globular blue berries are covered with bloom. They grow in clusters at the ends of the branches. The five calyx teeth are very prominent. Each berry contains many small seeds and is usually ten-celled. June, July.

Leaves. — The oval-lanceolate leaves are stemless and acute at both ends. The teeth are minute and bristle-like. Each surface is shining, but the lower one is lighter green than the upper. They are alternate in arrangement. In autumn they change to red colorings and fall early.

Flowers. — The white bell-shaped flowers grow in few-flowered racemes.

This is a dwarf shrub, with rough green branches, which are thickly covered with tiny white, raised dots. It is the earliest of the Blueberries to ripen, growing usually in rather exposed positions. It favors a thin, sandy soil, and especially frequents dry pine woods. It has a sweet and delicious flavor and such tiny seeds that it is a much more pleasant berry to eat than the Huckleberry. It is soft, however, and easily

bruised, which prevents its being largely marketed. George Emerson says it is suitable for
drying, and then forms a good substitute for
currants, for use in cakes, etc. The cluster of
ripening fruit presents an attractive color combination, with its green, pink, red, and blue berries.

Vaccinium Canadense, or Canadian Blueberry,
is similar to the preceding, but has leaves
which are downy on both sides and which
have entire margins. The branchlets are also
downy. The fruit ripens later in July or
August. It has a more northern range, being
most abundant in Canada. It is also found
along the mountains, south to Virginia. It
likes moist woods and swamps.

LOW BLUEBERRY

Vaccinium vacillans   **Huckleberry Family**

Fruit. — The berries of this shrub are borne
in raceme-like clusters at the end of a nearly
leafless twig. The calyx teeth are plainly visible at the summit. The fruit when ripe is
blue, with a bloom. It is slightly more acid
than *Vaccinium Pennsylvanicum* but of good

flavor. The berries ripen later than the preceding. July, September.

Leaves. — The leaves are oval or obovate, dull
green above and glaucous beneath. They are
narrowed or rounded at the base, and have
short stems. The apex is acute and ends in
a short bristle. The margin is entire or nearly
so, and the leaves are alternate. In the fall
the foliage changes to deep reds.

Flowers. — The pink or greenish bell-shaped
blossoms are somewhat contracted at the mouth.
They grow in clusters.

This shrub varies from one to four feet in
height. It is stiff and erect. It grows in light
soil along the wood borders and shaded roadsides from New Hampshire west to Michigan,
and south to Carolina and Missouri. The plant
is prolific, and when all the berries in the cluster
have ripened the fruit may be stripped off by
handfuls.

ARROWWOOD

Viburnum dentatum   **Honeysuckle Family**

Fruit. — The fruits are blue, — dark lead,
Emerson calls them, — but when gathered or

overripe become bluish black. They are oval, with calyx teeth and stigmas at the pointed tip. The flesh is thin, and the stone rounded on one side and with a rather deep groove on the other, making a cross section resemble a horse-shoe that has been flattened at the toe. The fruits grow in a flat-topped, erect cluster. They are dry and puckery, but are eaten by birds. August, September.

Leaves. — The leaves are opposite, coarsely and prominently toothed, ovate, pointed at the tip, and rounded or heart-shaped at the base. The petioles are short. Little tufts of hair are often in the axils of the midrib and branching veins on the lower surface. The leaves are yellowish green. Dark red is the autumnal color.

This shrub is from five to fifteen feet high, and has smooth gray bark. The under leaf surface has the little clusters of down in the axils. The name Arrowwood is applied to the shrub from the use made of the young shoots for arrows by the Indians. It inhabits moist places and borders streams. It extends south, along the mountains, to Georgia and west to Minnesota.

ARROWWOOD (*Viburnum dentatum*)

SOFT–LEAVED ARROWWOOD

Viburnum molle   **Honeysuckle Family**

Fruit. — The blue drupe is similar to the fruit of the preceding species. It is larger, sharply pointed, and oily. The depression of the stone is not so deep as in *Viburnum dentatum*.

Leaves. — The leaves are somewhat larger than those of the Arrowwood, but differ principally in being covered with soft hairs on the under surface.

Viburnum molle is principally distinguished from *Viburnum dentatum* by the pubescence on twigs, leaf, and flower stems, and lower leaf surfaces. It grows along the coast from eastern Massachusetts to Florida, and Texas.

BLUE OR MOUNTAIN FLY HONEYSUCKLE

Lonicera cœrulea   **Honeysuckle Family**

Fruit. — The berry is formed by the coalescence of two maturing ovaries. The exterior of the fruit shows its double structure by the two tiny " eyes " at the apex, each marking the rem-

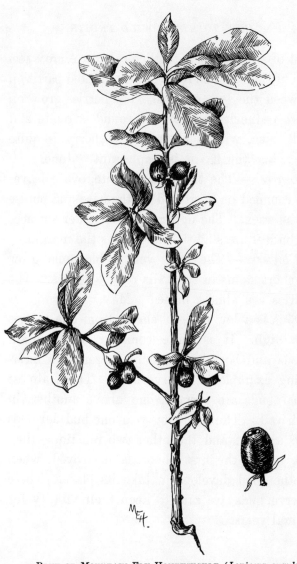

BLUE OR MOUNTAIN FLY HONEYSUCKLE (*Lonicera cœrulea*)

nant of a slightly five-toothed calyx. A cross section of the berry shows a clearly marked partition between the two ovaries. The berries grow on short peduncles. They are round or ovate and dark blue, with a bloom. The berry is quite juicy, but the flavor is unpleasant. June.

Leaves. — The thickish, opposite, ovate leaves are rounded or narrowed at the base and obtuse at the apex. The upper and the under surfaces of the leaf are slightly hairy, as is the margin.

Flowers. — The pale yellow blossoms grow on short stems in the axils of the leaves. The ovaries are almost united. May.

This is a low upright shrub, from one to two feet high. It is quite common in mountain woods and bogs. The plant presents an interesting example of reserve buds. Three almost equal buds are formed, one above another, in each axil. The following year one bud develops into a shoot and the other two remain as they are, unless the first shoot is destroyed, when another bud develops to take its place. These reserve buds are said to keep their vitality for several years.

YELLOW

YELLOW

NORTH AMERICAN PAPAW

Asimina triloba   **Custard-Apple Family**

Fruit. — Several large fleshy berries are borne together on a thickened peduncle. These fruit stems grow laterally from the axils of last year's leaves. Each berry is from two to six inches long and somewhat resembles a green banana. Its color, when ripe, is a yellowish green, and it is covered with a whitish bloom. The pulp is light yellow and of a fine grain, is soft and sweet. Two rows of flat beanlike seeds are arranged horizontally and alternate with each other throughout the length of the berry. The seeds are inclosed in fleshy arils. They are large, and form an obstacle to the pleasure of eating the fruit; the flavor is also too aromatic to be greatly relished. October.

Leaves. — The leaves are large, from ten to twelve inches long and four to five broad. They are entire, alternate, and short-petioled. They are reverse egg-shaped with acute apex, and pointed or slightly rounded base. The color of the fall leaf is dirty yellow.

Flowers. — The solitary flowers are green at opening, changing through browns and yellows to a deep red. They have two rows of petals; the outer three spreading, and the inner three erect, forming a sort of cup. April.

This is a low tree or shrub, forming a thick undergrowth in many forests, especially throughout the Mississippi valley. The foliage is dense and gives to the plant a tropical aspect. It is our one representative of a family which includes many tropical species. Rich, moist, woodland spots and banks of streams are the localities which it prefers. Its northern limit is Ontario and western New York. It extends west to Michigan and southward.

MAY APPLE. MANDRAKE. UMBRELLA LEAF. WILD LEMON

Podophyllum peltatum   **Barberry Family**

Fruit. — The large ovoid or lemon-shaped yellowish berry usually grows from the fork of two leaves. The fruit is fleshy and incloses numerous seeds, each of which is surrounded by a pulpy aril. These seeds are arranged in rows along a large lateral placenta. The fruit is sweet and edible. It retains the thickened stigma at the apex. July.

Leaves. — The leaves are five- to nine-lobed. The lobes are two-cleft and pointed at the apex. The flowerless stalks bear single leaves with the stems terminating near the center, giving the leaves a truly umbrella-like appearance. The leaves of the flowering stalks are in pairs, and their stems join the leaves nearer their inner edges. The upper surface is darker than the lower.

Flowers. — The large, white, drooping blossom, with its six to nine petals, is borne on a stout peduncle in the fork of the leaves. It is cross fertilized by bees, that visit the flowers for their pollen. They bear no nectar. April, May.

The leaves and the horizontal creeping root-stocks are poisonous if eaten, but possess certain medicinal properties. The plant spreads by means of its creeping rhizome and forms large patches. The umbrella-like leaf fulfills the mission suggested by its name for the pollen of the flower which it covers. The shape of the flower itself is evidently protective in a similar way. The manner in which the leaf forces its way through the ground is interesting. The lobes of the underground leaf are folded close to the stem, in closed umbrella fashion. At the point corresponding to the tip of the umbrella, the leaf cells are white and toughened, forming a hard knob. This, as the stem grows, bores its way through the earth to the surface. Above ground the group of cells softens, but remains as a white spot on the leaf. It grows in low woods, and is more prevalent in the Middle States than in New England.

DWARF THORN

Cratægus uniflora.   Cratægus parvifolia   Apple Family

Fruit. — The yellowish pome is globular or pear-shaped. It is usually solitary and borne on a short peduncle. The glandular, deeply cut calyx lobes are persistent.

Leaves. — The thick leaves are inversely egg-shaped. They are nearly stemless, and the upper portion of the margin has rounded teeth. The upper surface is shining and the lower one hairy.

Flowers. — The white flowers grow on short stems, usually alone, sometimes in pairs.

This low downy shrub favors sandy soil, and grows from southern New York south to Florida, where it reaches tree stature. It extends west to West Virginia and Louisiana. Both flower and fruit usually occur alone.

DWARF GINSENG. GROUNDNUT

Panax trifolium. Aralia trifolia   **Ginseng Family**

Fruit. — The berries are yellow, usually three-angled, but sometimes in united pairs. They are two- to three-seeded. They grow in a simple umbel.

Leaves. — The three compound leaves grow in a circle about the stem. There are three to five leaflets in each. The leaflets are sessile, the apex obtuse, the base narrowed, and the margin toothed.

Flowers. — The tiny white flowers grow in a small, fluffy, terminal cluster. April, May.

This is quite a common flower of our rich woods. It is seldom more than eight inches high. Its tuber is globular, edible, and aromatic, but is rather difficult to procure, being so deep in the ground. Georgia marks the southern limit.

DEERBERRY.   SQUAW HUCKLEBERRY

Vaccinium stamineum   **Huckleberry Family**

Fruit. — This berry is globose or pear-shaped, rather large, and greenish or yellowish. It is scarcely edible, falsely ten-celled, and few-seeded. The fruits grow in leafy-bracted racemes. September.

Leaves. — The oval or slightly heart-shaped leaves have short, downy petioles. They are whitened or slightly pubescent beneath, and the margins are slightly rolled backwards.

Flowers. — The flowers are distinguished by their long stamens, which project far beyond the short white corollas. The flowers grow in graceful clusters, with leaf bracts smaller than the regular leaves.

This is a much-branched shrub, from two to

five feet high. It grows in dry woods from
Maine and Minnesota south to Florida and
Louisiana.

PERSIMMON. DATE PLUM

Diospyros Virginiana   **Ebony Family**

Fruit. — The fruit is plumlike in appearance,
but botanically is a berry, with sometimes as
many as eight large flat seeds. When green, it
is very astringent, and in the north needs the
action of the frost to make it sweet and delicious. It then changes from a yellowish color
to a yellowish brown. The style is at the
summit, and the thick, four- to six-lobed calyx,
at the base. September, November.

Leaves. — The thickish leaves are dark green
above and paler beneath. They are nearly
smooth, ovate, pointed at the apex, and narrowed or rounded at the base.

Flowers. — The flowers are usually diœcious.
The fertile ones are solitary and grow in the
axils, while the smaller sterile ones are in small
clusters.

This tree is essentially southern, although it

occurs occasionally as far north as Rhode Island and New York. The fruit is used in beers and brandies. The wood is blackish in color and is well adapted for use in carving. Shoe lasts are made from it. The Duke of Argyle is said to have given a Persimmon Tree to King George III.

LOW HAIRY GROUND CHERRY

Physalis pubescens   **Potato Family**

Fruit. — This yellowish berry does not fill the small, short, membranous calyx. The berry is sticky. The solitary fruits grow from the leaf axils.

Leaves. — The leaves are entire or somewhat wavy and angled. They are rather small, from one inch to an inch and a half in length. They are pubescent or nearly smooth with hairy veins.

Flowers. — The flowers are yellow, with dark centers and purplish anthers.

This is an annual which is much-branched, and has pubescent stems and leaves. It favors sandy soil from New York to Minnesota, and south to Florida and Texas.

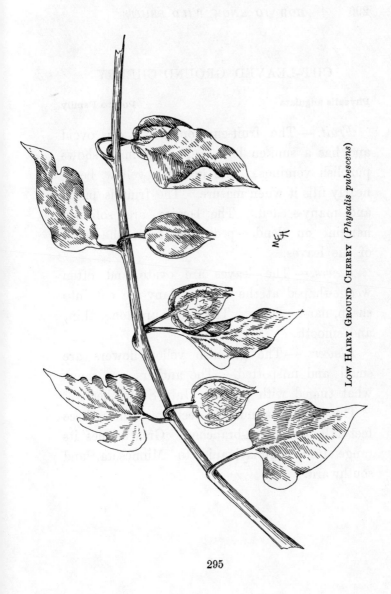

Low Hairy Ground Cherry (*Physalis pubescens*)

CUT–LEAVED GROUND CHERRY

Physalis angulata   **Potato Family**

Fruit. — The fruit-enveloping calyx is ovoid and has a sunken base. It sometimes shows purplish veinings. The greenish yellow berry nearly fills it when mature. The fruit is pulpy and many-seeded. The berries are solitary, hanging on slender peduncles from the axils of the leaves.

Leaves. — The leaves are ovate and often wedge-shaped at the base. Many are cut into sharp, narrow lobes. The leaves are long, thin, and smooth.

Flowers. — The greenish yellow flowers are small and unspotted. The anthers are somewhat tinged with purple. July–September.

This smooth annual is erect, sometimes three feet tall, and much-branched. Gray defines its range from Pennsylvania to Minnesota, and southward.

CLAMMY GROUND CHERRY

Physalis heterophylla   **Physalis Virginiana**

Potato Family

Fruit. — The yellow berry is loosely inclosed in the membranous calyx, which is much sunken at the stem. The fruit stem and calyx are pubescent. The solitary fruits hang along the branches from the leaf axils.

Leaves. — The leaves are broad, thick, and somewhat heart-shaped. They and the petioles are hairy. The apex is generally acute and the margin wavy, often having irregular, tooth-like lobes.

Flowers. — The yellow, five-lobed flowers have brown centers and yellow anthers.

This is the most common species and is very variable. It is a viscid, hairy, much-branched, spreading perennial. It extends from Ontario and Minnesota to Texas and Florida.

Var. ambigua is coarser, and coarsely covered with long soft hairs. The anthers are violet.

HORSE NETTLE

Solanum Carolinense   **Potato Family**

Fruit. — The smooth globular berries are orange-yellow and nearly three-quarters of an inch through. They grow in small, usually lateral clusters, on prickly stems. The calyx lobes are at the base of each berry. They are many-seeded.

Leaves. — The ovate or oblong leaves have wavy margins or are lobed with acute or obtuse lobes. The veins of the leaves are often prickly.

Flowers. — The violet flowers grow in terminal racemes, which become lateral in fruit.

This perennial of sandy waste places is hairy and has branches, stems, and parts of leaves thickly set with yellowish prickles. It extends from Connecticut west to Iowa and south.

GREEN

GREEN

GREEN ARROW–ARUM

Peltandra Virginica   **Peltandra undulata**

Arum Family

Fruit. — The berries grow in a head similar to the fruits of Indian Turnip and Dragon Root. The Green Arrow-arum berries, however, are green and nearly inclosed in the lower portion of the sheathing spathe. The upper part of the spathe breaks off before fruit develops. The one to three large seeds are inclosed in a colorless, jelly-like mass. The fruit-bearing stem is recurved, and the fruit bends to the water.

Leaves. — The leaves are shaped like long arrow heads. One prominent vein extends from base toward tip and one from base into each basal lobe.

Flowers. — The tapering spadix is covered with flowers throughout its length. It bears both staminate and pistillate flowers, the latter

at its base.   The spathe incloses the entire
length of the spadix, with the exception of an
oval opening in front, about midway of its
length.

Clusters of Green Arrow-arum grow in shal-
low water along river borders.   It is about a
foot and a half high.   The sheathed fruit looks
much like a pond lily bud and bends on its
recurved stem to the surface of the water.   It is
quite possible that the fruit heads are broken by
the force of the water and carried down stream
to originate new colonies.   Lack of bright
coloring in the fruit is suggestive of some means
of seed dispersal other than by agency of birds.
The plant extends west to Michigan and Louisiana.

CHOKEPEAR

Pyrus communis   **Apple Family**

The Chokepear Tree, with its green, puckery
pear, hardly needs description.   I trust many
another like myself holds it in grateful remembrance for the childhood joys it has furnished.
What an addition the fruit was, on chestnutting
expeditions, to the ginger cookies, which always

started out in the pail with us never to return. If you have never eaten it cooked as a sauce and flavored with molasses, you have yet to taste a very delectable sirup.

AMERICAN CRAB APPLE

Malus coronaria   **Pyrus coronaria**

Apple Family

Fruit. — The apple is yellowish green and flattened lengthwise. The small, smooth calyx lobes are in the deep, broad depression at the summit. The fruit stem is slender. One or two dark brown seeds are in each cell. The fruit is fragrant but sour. It hangs long on the trees, and does not usually decay until the following spring.

Leaves. — The leaves are ovate or triangular-ovate, they grow on short, slender, hard petioles. The margins are serrate. Sometimes the leaf is three-lobed. It is yellow in autumn.

Flowers. — The flowers are much like those of the cultivated apple, but very fragrant and of a beautiful pink color.

This tree is often planted near homes, because

of the fragrance and beauty of the blossoms.
The fruit is sometimes used for cider. The tree
is not very high, spreading, and often thorny.
This is particularly a northern species.

A similar tree, *Malus angustifolia*, is the
southern species, although the two overlap in
range. Bailey says that the best mark of
distinction between the two " is the thick, half-
evergreen shining leaves of *Malus angustifolia*."
The flowers are smaller than in preceding
species.

WHITE

WHITE

BAYBERRY. WAXBERRY

Myrica Carolinensis   **Myrica cerifera**

Bayberry Family

Fruit. — The fertile aments develop into clusters of dry drupes, with from four to nine separate fruits in the cluster. These clusters are fastened to the branches by short stalks. Each drupe is covered with many tiny grains, which finally become coated with white wax. The covering is first green, then blackish, and finally white. The stone is hard. The fruits persist for two or three years.

Leaves. — The obovate or oblanceolate leaves are nearly stemless. They have resinous dots on both sides, are leathery, shining, bright green, and aromatic. The margin is slightly toothed toward the apex, otherwise entire. The base is narrowed and the apex obtuse, sometimes acute, or often ends abruptly in a sharp point.

Flowers. — The fertile catkins are small and erect, and consist of several ovaries, which are sheltered by scales. April–June.

The fruits in the earlier days of our country's history were much prized for the wax which they yield. This is obtained by boiling the drupes and skimming the wax from the surface of the water. It was used for making candles, either alone or mixed with tallow or beeswax. The Bayberry candles emit a pleasant odor, but their light is not so bright as the flame of the tallow candles. The Bayberry will grow in almost any soil. It extends along the eastern coast, and occurs somewhat near the Great Lakes.

WHITE MULBERRY

Morus alba   **Mulberry Family**

Fruit. — The structure of the fruit is the same as that of the Red Mulberry. The fruit of *Morus alba*, however, is white, shorter, and not as juicy. July.

Leaves. — The shining dark green leaves are variable in shape, serrate, and shining.

The White Mulberry Tree grows rapidly, reaching a height of thirty or forty feet. It is a native of China, and its leaves are extensively used as food for silkworms. The tree was introduced into America when silkworm raising was being tried in this country, and occurs now spontaneously near houses, especially in the vicinity of long-established silk manufacturing plants.

SMALL MISTLETOE

Razoumofskya pusilla   **Arceuthobium pusillum**

Mistletoe Family

Fruit. — The ovoid-oblong berries are solitary and grow on short recurved stems. They are fleshy, with seeds inclosed in a sticky mucus. They develop in the autumn, a year or more after flowering.

This is an inconspicuous parasite, drawing its nourishment from branches of the fir. It is olive green to brown in color, and the leaves are obtuse and scale-like.

AMERICAN MISTLETOE

Phoradendron flavescens   **Mistletoe Family**

Fruit. — The white berries are globose, pulpy, and one-seeded. They grow in clusters on a short foot stalk.

Leaves. — The leaves are thick, leathery, yellowish green, oval or obovate, entire, obtuse at apex and narrowed into a short petiole at the base. They are persistent throughout the season.

Flowers. — The dioecious flowers grow in catkinlike spikes. May–July.

This parasite flourishes on deciduous trees, notably the Red Maple and Tupelo. Its wood is yellowish green, and the thick, firm leaves and white berries persist during the winter. The Mistletoe has a place in Christmas decorations, and may often be seen at that time exposed for sale. *Phoradendron* means tree-thief, referring to its parasitic life. While essentially southern, it occurs in southern New Jersey, Pennsylvania, Ohio, Indiana, and Illinois.

White Baneberry (*Actæa alba*).

381

WHITE BANEBERRY (*Actæa alba*)

312

WHITE BANEBERRY

Actæa alba   **Crowfoot Family**

Fruit. — The terminal fruit clusters of the White Baneberry are oblong and usually more open than those of the Red. The white berries are almost globular, have a black mark at the tip and a crease on one side, extending from the apex to the pedicel. The pedicel is thickened and usually red. The lower pedicels are much longer than the upper ones. The numerous large brown seeds are packed horizontally. A plant with red berries on thickened red stalks sometimes occurs. The fruit develops in August, about a month later than the Red Baneberry, and persists into September.

Leaves. — The leaves are twice or thrice compound, with deeply cut, acute lobes and sharp teeth.

Flowers. — The petals are so like stamens as to seem to be transformed stamens. The flowers yield no honey, simply pollen, to the bees, which secure their cross fertilization.

This herb grows in woods as far south as Pennsylvania and New Jersey.

POISON SUMAC

Rhus Vernix.   Rhus venenata   **Sumac Family**

Fruit. — The smooth, somewhat glossy drupe is grayish. It is dry and slightly pear-shaped, with the sides unequal. It grows in open loose clusters from the leaf axils. It closely resembles the fruit of the Poison Ivy. August, September.

Leaves. — The stalks of the compound leaves are usually purplish. There are from three to thirteen nearly stemless leaflets, which are unequal at the base. They are a bright shining green, acute at the apex, *entire*, and obovate or oval. The autumnal colorings are most brilliant.

Flowers. — The small, greenish yellow, diœcious flowers grow in open loose panicles from the leaf axils.

It is not strange that many an unfortunate, hoping to prolong his enjoyment of the brilliant foliage, should be lured into gathering its autumnal leaves for home decorations. Immune, he may be, it is true, but doubtless a long period of suffering will follow his rash act.

Some persons are poisoned by even passing near the plant, contact not being necessary.

If in fruit, the whitish color of the drupes and their drooping clusters are sure marks by which this Sumac may be distinguished from the other species. The entire leaves and lack of winged petioles and pubescence are also marks of distinction.

POISON, CLIMBING, OR THREE–LEAVED IVY

Rhus radicans   **Rhus Toxicodendron**

Sumac Family

Fruit. — The fruit closely resembles that of Poison Sumac. September and persistent.

Leaves. — The compound leaves have *three* pale green leaflets, which are sharply toothed and entire or sometimes lobed.

This plant is sometimes erect and one to three feet high, sometimes prostrate and trailing, and sometimes climbing. It supports itself by numerous rootlets, which penetrate and hold tenaciously to various supports. Its three-parted leaves and white fruit distinguish this poisonous

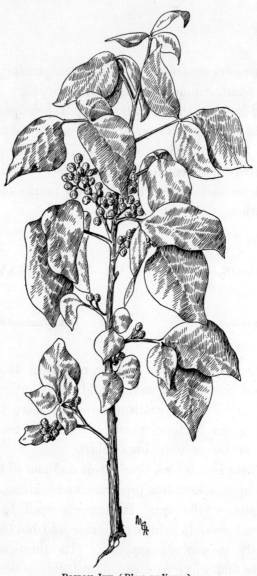

POISON IVY (*Rhus radicans*)

plant from the harmless Virginia Creeper, *Parthenocissus quinquefolia*, which is somewhat similar in its manner of growth. The dryish fruits are used as food by the winter birds. For crows especially they serve as an important article of diet. One hundred and fifty-three Poison Ivy seeds are said to have been found in the stomach of one of these birds. The dry outer husks are removed by action of stomach and thrown out again in small masses through the mouth.

RED–OSIER CORNEL OR DOGWOOD

Cornus stolonifera   **Dogwood Family**

Fruit. — The drupe is white, or whitish, and globose. The stone is very variable in shape. The fruits grow in flat-topped, rather smallish cymes.

Leaves. — The ovate or ovate-lanceolate leaf has an abrupt, short, tapering apex and a rounded base. The upper surface is finely pubescent and the lower whitish and somewhat downy.

Flowers. — The smallish flat cymes are rather few-flowered. June, July.

Reddish branches are a characteristic feature

of this dogwood. They are especially brilliant in late winter and early spring. The main stem is usually prostrate, often unnoticed because of a covering of leaves. This sends down rootlets and sends up slender branches, soon forming broad clumps. The main shoot is sometimes underground.

The range of the shrub is from ocean to ocean, extending south to Virginia, Kentucky, Nebraska, Arizona, and California. A similar species occurs in Siberia.

PANICLED CORNEL

Cornus candidissima   **Cornus paniculata**
Dogwood Family

Fruit. — The small, white, flattened, globose drupes grow in convex clusters. The peduncle and pedicels are red. The plant often fruits sparingly, and the clusters are consequently ragged and irregular. Each fruit is crowned with minute calyx teeth, through which the style protrudes. The flesh is thin and white, inclosing a two-celled, two-seeded stone. August, September. This is one of the earliest fruited Dogwoods.

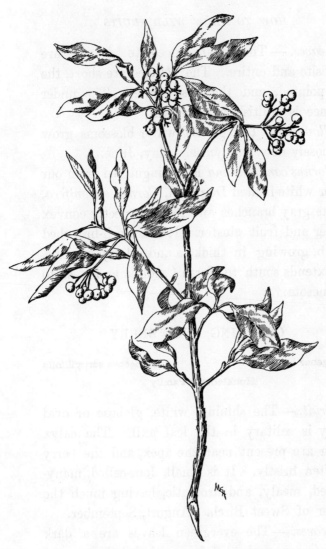

PANICLED CORNEL (*Cornus candidissima*)

319

Leaves. — The ovate-lanceolate leaves are opposite and entire. The petioles are short, the tip pointed, and the base acute. The under surface is whitish but smooth.

Flowers. — The perfect white blossoms grow in loosely flowered cymes. May, June.

Cornus candidissima is distinguished from our other white-fruited Dogwood, *Cornus stolonifera*, by its gray branches and its imperfectly convex flower and fruit cluster. It is a much-branched shrub, growing in thickets and along streams. It extends south to North Carolina and west to Minnesota.

CREEPING SNOWBERRY

Chiogenes hispidula   **Chiogenes serpyllifolia**
Huckleberry Family

Fruit. — The shining, white, globose or oval berry is solitary in the leaf axil. The calyx teeth are present near the apex, and the berry is often bristly. It is small, four-celled, many-seeded, mealy, and aromatic, having much the flavor of Sweet Birch. August, September.

Leaves. — The evergreen leaves are a dark olive green, with stiff brownish bristles on the

SNOWBERRY (*Symphoricarpos racemosus*)
322

under surface. They are small, on short stems, and have backward rolled margins.

Flowers. — The tiny white nodding flowers are single, growing on short stems from the leaf axil. Two bracts are beneath the calyx. May.

The stems of this creeping and trailing shrub are scarcely woody, slender, and bristly. It is a native of Japan, and in our country extends from Newfoundland to British Columbia and south to Michigan, New Jersey, and Pennsylvania, continuing along the Alleghanies to North Carolina. *Chiogenes* means "snow offering," referring most appropriately to the snow-white fruit.

SNOWBERRY

Symphoricarpos racemosus   **Honeysuckle Family**

Fruit. — The terminal fruit spike usually has a pair of leaves at its base. Solitary berries sometimes grow from the axils of the next lower pair of leaves. The ripe berries are quite large, waxy, and white. The persistent calyx teeth are at the top and two tiny bracts are at the base. The berries are nearly stemless. They have two large cells, each containing a seed; and two smaller empty cells. The seed coats are hard. The berries begin to ripen in July, and the spikes show fruit in various stages of development, with buds and flowers at the summit well into September.

Leaves. — The ovate, usually entire leaves grow in pairs. The stems are quite short. The leaves are dark green above and lighter beneath.

Flowers. — The small bell-shaped pink blossom is four- or five-toothed. It is hairy at the throat.

This is a common inhabitant of old-fashioned gardens, and lingers about abandoned farmhouses or even the cellars. It strays beyond the garden bounds and often occurs along the roadsides. It grows also along rocky banks in New England to Pennsylvania and westward. It is most attractive in September, when the spike is nearly full of matured fruits.

Symphoricarpos pauciflorus appears in the mountains of Vermont and Pennsylvania and westward. The leaves are smaller than the preceding, and the berries grow singly or in pairs in the uppermost leaf axils.

GLOSSARY

Achene. A small, dry, indehiscent, one-seeded fruit with usually a thin pericarp.

Acute. Sharp-pointed.

Alternate. As opposed to opposite.

Ament. Synonymous with catkin. A spike consisting of imperfect flowers with a scale-like bract at the base of each.

Annual. Of but one season's growth.

Anther. The part of the stamen which yields the pollen.

Aril. A fleshy seed covering growing from the cord which attaches the seed to the seed vessel.

Axil. The angle which the stem forms with a leaf or branch.

Bloom. A secretion of wax covering the surface of leaf or fruit.

Bract. A modified leaf at the base of flower or fruit.

Bristle. A stiff, hairlike growth.

Calyx. The outer and protective floral whorl.

Capsule. A dry, dehiscent fruit of two or more carpels.

Carpel. This is the seed-bearing part of the flower. It may be a simple pistil or one of the parts of a compound pistil.

Catkin. Same as ament.

Cell. A structure inclosing a cavity.

Ciliate. With hairy margin.

Compound. A whole made up of two or more similar parts.

Connate. The joining of similar organs. Opposite leaves whose bases join.

Corm. Like a solid bulb.

Corolla. The inner floral whorl.

Corymb. A flat-topped or rounded flower cluster with the marginal flowers opening first.

Cotyledon. A seed leaf or leaves.

Cross Fertilization. The action resulting from the deposit of the pollen of one flower on the stigma of another.

Cyme. A flattish flower cluster with the blossoms unfolding from the center outwards.

Deciduous. Not persistent.

Dicotyledonous. With two cotyledons.

Diœcious. Staminate and pistillate flowers borne on different plants of the same species.

Disk. A thickened circle of cellular tissue about the base of the stamens and around the ovary.

Fertile. Productive.

Gland. A secreting surface or structure.

Glaucous. Covered with a bloom.

Herb. With the above-ground stems living but one season.

Involucre. Bracts surrounding a single flower or a flower cluster or head.

Lanceolate. Much longer than broad. The widest portion is below the middle, and the leaf tapers towards either end.

Monocotyledonous. With but one cotyledon.

Monœcious. Separate staminate and pistillate flowers on the same plant.

Node. The joints on the stem where a leaf or whorl of leaves would naturally grow.

Oblanceolate. The broadest portion of the long leaf nearest the apex and tapering to either end.

Oblong. Longer than broad, and with sides nearly parallel.

Obovate. Egg-shaped, with broader portion nearest apex.

Obtuse. With a rounded or blunt end.

Ovary. The part of the pistil containing the ovules.

Ovule. The body which, when fertilized, is meant to become a seed.

Panicle. A loose, irregular cluster with branching flower stems.

Parasitic. Gaining nourishment from a host plant.

Pedicel. The stem of one of the component flowers or fruits of a cluster.

Peduncle. A stem of a single flower or of a flower cluster.

Perennial. Continuing year after year.

Perfect (flower). Having both stamens and pistils.

Perianth. The floral envelope consisting of calyx and corolla if both are present.

Petal. One unit of the corolla.

Petiole. The leaf stalk.

Pinnate. With the leaflets of a compound leaf on either side of the leaf stalk.

Placenta. The interior portion of the ovary on which the ovules are borne.

Pollen. The anther-borne grains which fertilize the ovules.

Polygamous. Plants bearing staminate, pistillate, and perfect flowers on the same plant.

Prickles. A slender, sharp growth from bark or rind.

Pubescent. Finely hairy.

Raceme. Spike bearing stemmed flowers.

Receptacle. The modified portion of an axis upon which the flowers or portion of a flower is borne.

Rhizome. An underground stem.

Root. The underground part of the plant which obtains nourishment there.

Scape. A naked flower stem springing from the ground.

Seed. The ripened ovule.

Sepal. One unit of the calyx.

Serrate. With forward-pointing teeth.

Sessile. Stemless.

Shrub. Plants with woody structure and with several stems springing from the ground or near it, or whose stems are much-branched. Usually smaller than trees.

Sinus. The margin between the lobes.

Spadix. A spike having a fleshy axis.

Spathe. One or more large bracts inclosing an inflorescence.

Spike. Sessile or nearly sessile flowers borne on a somewhat elongated axis.

Spine. A sharp growth from the stem.

Sterile. Unproductive.

Stigma. The portion of the pistil receptive to the pollen grain.

Stipule. An appendage at the base of a leaf stem, sometimes joined to the petiole.

Style. The portion of the pistil connecting the stigma and ovary.

Tendril. A slender, coiling part of a climbing plant, aiding in its support.

Umbel. A flower cluster in which the pedicels spring from a common point.

Whorl. A circular arrangement of leaves, etc., around a stem.

ABBREVIATIONS OF AUTHORS' NAMES

Ait., Aiton.
Andr., Andrews.
C. & S., Chamisso and Schlechtendahl.
DC., De Candolle.
Desf., Desfontaine.
Desv., Desvaux.
Dietr., Dietrich.
Ehrh., Ehrhart.
Ell., Elliott.
Hook., Hooker.
Jacq., Jacquin.
Karst., Karsten.
L., Linnæus.
Lam., Lamarck.
L. f., Linne (the son).
L'Her., L'Heritier de Brutelle.
Lodd., Loddiges.
MacM., MacMillan.
Marsh., Marshall.

Medic., Medicus.
Michx., Michaux.
Mill., Miller.
Muench., Muenchhausen.
Muhl., Mühlenberg.
Nutt., Nuttall.
Pers., Persoon.
Planch., Planchon.
Poir., Poiret.
R. & S., Roemer and Schultes.
Roem., Roemer.
Salisb., Salisbury.
Spreng., Sprengel.
Sudw., Sudworth.
Torr., Torrey.
T. & G., Torrey & Gray.
Vent., Ventenat.
Walt., Walter.
Wang., Wangenheim.
Willd., Willdenow.

INDEX TO ENGLISH NAMES

INDEX TO LATIN NAMES

A CATALOGUE OF SELECTED DOVER BOOKS
IN ALL FIELDS OF INTEREST

A CATALOGUE OF SELECTED DOVER BOOKS
IN ALL FIELDS OF INTEREST

LEATHER TOOLING AND CARVING, Chris H. Groneman. One of few books concentrating on tooling and carving, with complete instructions and grid designs for 39 projects ranging from bookmarks to bags. 148 illustrations. 111pp. 7⅞ x 10.
23061-9 Pa. $2.50

THE CODEX NUTTALL, A PICTURE MANUSCRIPT FROM ANCIENT MEXICO, as first edited by Zelia Nuttall. Only inexpensive edition, in full color, of a pre-Columbian Mexican (Mixtec) book. 88 color plates show kings, gods, heroes, temples, sacrifices. New explanatory, historical introduction by Arthur G. Miller. 96pp. 11⅜ x 8½.
23168-2 Pa. $7.50

AMERICAN PRIMITIVE PAINTING, Jean Lipman. Classic collection of an enduring American tradition. 109 plates, 8 in full color—portraits, landscapes, Biblical and historical scenes, etc., showing family groups, farm life, and so on. 80pp. of lucid text. 8⅜ x 11¼.
22815-0 Pa. $4.00

WILL BRADLEY: HIS GRAPHIC ART, edited by Clarence P. Hornung. Striking collection of work by foremost practitioner of Art Nouveau in America: posters, cover designs, sample pages, advertisements, other illustrations. 97 plates, including 8 in full color and 19 in two colors. 97pp. 9⅜ x 12¼.
20701-3 Pa. $4.00
22120-2 Clothbd. $10.00

THE UNDERGROUND SKETCHBOOK OF JAN FAUST, Jan Faust. 101 bitter, horrifying, black-humorous, penetrating sketches on sex, war, greed, various liberations, etc. Sometimes sexual, but not pornographic. Not for prudish. 101pp. 6½ x 9¼.
22740-5 Pa. $1.50

THE GIBSON GIRL AND HER AMERICA, Charles Dana Gibson. 155 finest drawings of effervescent world of 1900-1910: the Gibson Girl and her loves, amusements, adventures, Mr. Pipp, etc. Selected by E. Gillon; introduction by Henry Pitz. 144pp. 8¼ x 11⅜.
21986-0 Pa. $3.50

STAINED GLASS CRAFT, J.A.F. Divine, G. Blachford. One of the very few books that tell the beginner exactly what he needs to know: planning cuts, making shapes, avoiding design weaknesses, fitting glass, etc. 93 illustrations. 115pp.
22812-6 Pa. $1.50

MANUAL OF THE TREES OF NORTH AMERICA, Charles S. Sargent. The basic survey of every native tree and tree-like shrub, 717 species in all. Extremely full descriptions, information on habitat, growth, locales, economics, etc. Necessary to every serious tree lover. Over 100 finding keys. 783 illustrations. Total of 986pp.
20277-1, 20278-X Pa., Two vol. set $8.00

BIRDS OF THE NEW YORK AREA, John Bull. Indispensable guide to more than 400 species within a hundred-mile radius of Manhattan. Information on range, status, breeding, migration, distribution trends, etc. Foreword by Roger Tory Peterson. 17 drawings; maps. 540pp.
23222-0 Pa. $6.00

THE SEA-BEACH AT EBB-TIDE, Augusta Foote Arnold. Identify hundreds of marine plants and animals: algae, seaweeds, squids, crabs, corals, etc. Descriptions cover food, life cycle, size, shape, habitat. Over 600 drawings. 490pp.
21949-6 Pa. $4.00

THE MOTH BOOK, William J. Holland. Identify more than 2,000 moths of North America. General information, precise species descriptions. 623 illustrations plus 48 color plates show almost all species, full size. 1968 edition. Still the basic book. Total of 551pp. 6½ x 9¼.
21948-8 Pa. $6.00

AN INTRODUCTION TO THE REPTILES AND AMPHIBIANS OF THE UNITED STATES, Percy A. Morris. All lizards, crocodiles, turtles, snakes, toads, frogs; life history, identification, habits, suitability as pets, etc. Non-technical, but sound and broad. 130 photos. 253pp.
22982-3 Pa. $3.00

OLD NEW YORK IN EARLY PHOTOGRAPHS, edited by Mary Black. Your only chance to see New York City as it was 1853-1906, through 196 wonderful photographs from N.Y. Historical Society. Great Blizzard, Lincoln's funeral procession, great buildings. 228pp. 9 x 12.
22907-6 Pa. $6.00

THE AMERICAN REVOLUTION, A PICTURE SOURCEBOOK, John Grafton. Wonderful Bicentennial picture source, with 411 illustrations (contemporary and 19th century) showing battles, personalities, maps, events, flags, posters, soldier's life, ships, etc. all captioned and explained. A wonderful browsing book, supplement to other historical reading. 160pp. 9 x 12.
23226-3 Pa. $4.00

PERSONAL NARRATIVE OF A PILGRIMAGE TO AL-MADINAH AND MECCAH, Richard Burton. Great travel classic by remarkably colorful personality. Burton, disguised as a Moroccan, visited sacred shrines of Islam, narrowly escaping death. Wonderful observations of Islamic life, customs, personalities. 47 illustrations. Total of 959pp.
21217-3, 21218-1 Pa., Two vol. set $7.00

INCIDENTS OF TRAVEL IN CENTRAL AMERICA, CHIAPAS, AND YUCATAN, John L. Stephens. Almost single-handed discovery of Maya culture; exploration of ruined cities, monuments, temples; customs of Indians. 115 drawings. 892pp.
22404-X, 22405-8 Pa., Two vol. set $8.00

DRIED FLOWERS, Sarah Whitlock and Martha Rankin. Concise, clear, practical guide to dehydration, glycerinizing, pressing plant material, and more. Covers use of silica gel. 12 drawings. Originally titled "New Techniques with Dried Flowers." 32pp. 21802-3 Pa. $1.00

ABC OF POULTRY RAISING, J.H. Florea. Poultry expert, editor tells how to raise chickens on home or small business basis. Breeds, feeding, housing, laying, etc. Very concrete, practical. 50 illustrations. 256pp. 23201-8 Pa. $3.00

HOW INDIANS USE WILD PLANTS FOR FOOD, MEDICINE & CRAFTS, Frances Densmore. Smithsonian, Bureau of American Ethnology report presents wealth of material on nearly 200 plants used by Chippewas of Minnesota and Wisconsin. 33 plates plus 122pp. of text. 6⅛ x 9¼. 23019-8 Pa. $2.50

THE HERBAL OR GENERAL HISTORY OF PLANTS, John Gerard. The 1633 edition revised and enlarged by Thomas Johnson. Containing almost 2850 plant descriptions and 2705 superb illustrations, Gerard's Herbal is a monumental work, the book all modern English herbals are derived from, and the one herbal every serious enthusiast should have in its entirety. Original editions are worth perhaps $750. 1678pp. 8½ x 12¼. 23147-X Clothbd. $50.00

A MODERN HERBAL, Margaret Grieve. Much the fullest, most exact, most useful compilation of herbal material. Gigantic alphabetical encyclopedia, from aconite to zedoary, gives botanical information, medical properties, folklore, economic uses, and much else. Indispensable to serious reader. 161 illustrations. 888pp. 6½ x 9¼. USO 22798-7, 22799-5 Pa., Two vol. set $10.00

HOW TO KNOW THE FERNS, Frances T. Parsons. Delightful classic. Identification, fern lore, for Eastern and Central U.S.A. Has introduced thousands to interesting life form. 99 illustrations. 215pp. 20740-4 Pa. $2.50

THE MUSHROOM HANDBOOK, Louis C.C. Krieger. Still the best popular handbook. Full descriptions of 259 species, extremely thorough text, habitats, luminescence, poisons, folklore, etc. 32 color plates; 126 other illustrations. 560pp. 21861-9 Pa. $4.50

HOW TO KNOW THE WILD FRUITS, Maude G. Peterson. Classic guide covers nearly 200 trees, shrubs, smaller plants of the U.S. arranged by color of fruit and then by family. Full text provides names, descriptions, edibility, uses. 80 illustrations. 400pp. 22943-2 Pa. $3.00

COMMON WEEDS OF THE UNITED STATES, U.S. Department of Agriculture. Covers 220 important weeds with illustration, maps, botanical information, plant lore for each. Over 225 illustrations. 463pp. 6⅛ x 9¼. 20504-5 Pa. $4.50

HOW TO KNOW THE WILD FLOWERS, Mrs. William S. Dana. Still best popular book for East and Central USA. Over 500 plants easily identified, with plant lore; arranged according to color and flowering time. 174 plates. 459pp. 20332-8 Pa. $3.50

BUILD YOUR OWN LOW-COST HOME, L.O. Anderson, H.F. Zornig. U.S. Dept. of Agriculture sets of plans, full, detailed, for 11 houses: A-Frame, circular, conventional. Also construction manual. Save hundreds of dollars. 204pp. 11 x 16.
21525-3 Pa. $5.95

HOW TO BUILD A WOOD-FRAME HOUSE, L.O. Anderson. Comprehensive, easy to follow U.S. Government manual: placement, foundations, framing, sheathing, roof, insulation, plaster, finishing — almost everything else. 179 illustrations. 223pp. 7⅞ x 10¾.
22954-8 Pa. $3.50

CONCRETE, MASONRY AND BRICKWORK, U.S. Department of the Army. Practical handbook for the home owner and small builder, manual contains basic principles, techniques, and important background information on construction with concrete, concrete blocks, and brick. 177 figures, 37 tables. 200pp. 6½ x 9¼.
23203-4 Pa. $4.00

THE STANDARD BOOK OF QUILT MAKING AND COLLECTING, Marguerite Ickis. Full information, full-sized patterns for making 46 traditional quilts, also 150 other patterns. Quilted cloths, lamé, satin quilts, etc. 483 illustrations. 273pp. 6⅞ x 9⅝.
20582-7 Pa. $3.50

101 PATCHWORK PATTERNS, Ruby S. McKim. 101 beautiful, immediately useable patterns, full-size, modern and traditional. Also general information, estimating, quilt lore. 124pp. 7⅞ x 10¾.
20773-0 Pa. $2.50

KNIT YOUR OWN NORWEGIAN SWEATERS, Dale Yarn Company. Complete instructions for 50 authentic sweaters, hats, mittens, gloves, caps, etc. Thoroughly modern designs that command high prices in stores. 24 patterns, 24 color photographs. Nearly 100 charts and other illustrations. 58pp. 8⅜ x 11¼.
23031-7 Pa. $2.50

IRON-ON TRANSFER PATTERNS FOR CREWEL AND EMBROIDERY FROM EARLY AMERICAN SOURCES, edited by Rita Weiss. 75 designs, borders, alphabets, from traditional American sources printed on translucent paper in transfer ink. Reuseable. Instructions. Test patterns. 24pp. 8¼ x 11.
23162-3 Pa. $1.50

AMERICAN INDIAN NEEDLEPOINT DESIGNS FOR PILLOWS, BELTS, HANDBAGS AND OTHER PROJECTS, Roslyn Epstein. 37 authentic American Indian designs adapted for modern needlepoint projects. Grid backing makes designs easily transferable to canvas. 48pp. 8¼ x 11.
22973-4 Pa. $1.50

CHARTED FOLK DESIGNS FOR CROSS-STITCH EMBROIDERY, Maria Foris & Andreas Foris. 278 charted folk designs, most in 2 colors, from Danube region: florals, fantastic beasts, geometrics, traditional symbols, more. Border and central patterns. 77pp. 8¼ x 11.
USO 23191-7 Pa. $2.00

Prices subject to change without notice.
Available at your book dealer or write for free catalogue to Dept. GI, Dover Publications, Inc., 180 Varick St., N.Y., N.Y. 10014. Dover publishes more than 150 books each year on science, elementary and advanced mathematics, biology, music, art, literary history, social sciences and other areas.